AN ESCORT OF P-38s

The First Fighter Group in World War II

BY

JOHN D. MULLINS

PHALANX
Publishing Co., Ltd.
1051 Marie Avenue
St. Paul, MN 55118 U.S.A.
612/454-0607

ISBN: 1-883809-03-7

Library of Congress Number: 94-065588

Text by: John Mullins

Edited by: John Lambert

Cover Illustration and aircraft profiles by: John Valo

Maps by: Steven Carley

Printed in Hong Kong

Published by:

Phalanx Publishing Co., Ltd.
1051 Marie Avenue West
St. Paul, MN 55118-4131 USA

LIGHTNINGS IN THE SKY

By Technical Sergeant Robert Bryson*

Oh, Hedy Lamarr is a beautiful gal
And Madeleine Carroll is too;
But you'll find, if you query, a different theory
Amongst any bomber crew.
For the loveliest thing of which one could sing
(this side of the heavenly gates)
Is no blond or brunette of the Hollywood set-
But an escort of P-38's

Yes in days that have passed
When the tables were massed
With glasses of Scotch and Champagne,
It's quite true that the sight was a thing to delight
Us-intent on feeling no pain.
But no longer the same; nowadays in this game,
When we head north from Messina Straits,
Take the sparkling wine – everytime – just make mine
An escort of P-38's

Byron, Shelley and Keats ran a dozen dead heats
Describing the view from the hills;
Of the valleys in May when the winds gently sway
An army of bright daffodils.
Take the daffodils Byron – the wildflowers Shelley,
Yours is the myrtle friend Keats;
Just reserve me those cuties, those American Beauties
An escort of P-38's

Sure we're braver than Hell, on the ground all is swell;
In the air it's a different story;
We sweat out our track through the fighters and flak;
We're willing to split up the glory.
Well they wouldn't reject us so Heaven protect us;
And until all this shooting abates,
Give us courage to fight 'em and one other small item-
An escort of P-38's

* A crewman on a B-17 of the 429th Squadron, 2nd Bomb Group, Bryson was killed over Genoa, Italy in October 1943

TABLE OF CONTENTS

PREFACE

The history of the 1st Fighter Group is an outgrowth of a history, written earlier, of the 94th squadron, one component of the Group. With one squadron history completed the Group history did not appear to be quite so daunting an undertaking. Nonetheless, reviewing and compiling a history based on the microfilmed records of long ago achievements is not particularly interesting even if one had once been a part of those achievements. What makes a history interesting, even to more than the immediate participants, are the personal recollections, diaries, photographs and similar artifacts that make a written account more realistically represent living, breathing, human beings who were once caught up in a war experience beyond their wildest imaginings. The individual contributors have made this history come alive, once again to the author and, with any good fortune, to the interested reader as well.

Using the official records and adding to them personal recollections and experiences has been an humbling task and a good illustration of the risks involved in trusting memory. The passage of time draws a veil over our most vivid memories. Our experiences become fragmented and blurred. One of my friends, on reading the 94th history, remarked that he could remember each of the locations given for the squadron but not at all in the sequence listed. My own recollections have also been considerably rearranged when confronted by the official records. Still, by combining the two, the result is a far more readable and interesting yet accurate account of both major and minor events that comprise the history of the 1st Fighter Group and its supporting cast during the war years.

INTRODUCTION

This history describes the activities of the 1st Fighter Group during the years of the second world war. The group exists today as the 1st Fighter Wing. The Group, the oldest in the Air Corps, includes three squadrons, the 27th, 71st and 94th. The 27th and 94th date from the earliest days of the first war and, in fact predate the Group. The 71st was organized in 1941. The Group pilots flew the P-38 Lightning fighter aircraft throughout the war. The 1st flew the Atlantic to England (the Bolero Mission), became operational as a part of the fledgling 8th Air Force with 1942 forays over occupied France, took part in the war throughout North Africa and Sicily in the 12th Air Force and ended the war in Italy as a part of the 15th Air Force.

The history is based on the official records of the Group and the three squadrons much, but not all of which are on microfilm. Squadron and Group mission reports, for example, are stored in cardboard boxes in the National Archives building in Suitland, Md. Microfilmed records, available from the USAF Historical Research Center, Maxwell AFB, are a) frequently illegible and b) extremely terse and dry - an entire mission of 12 P-38's described in two sentences. For these reasons this history is supplemented by personal recollections, anecdotes, diary entries and photos to make the dust dry facts take on some color and personality.

Due, at least in part, to the records available, this history is one of operations. There is little that concerns the logistics or administrative details of either the squadrons or the group. We cannot recite the trials and tribulations of the cooks, for example, although their challenges must have been severe. The S-2 or Intelligence function is dealt with as is Armament and some of the support functions, mechanics and the like, thanks to personal contributions. Medical records are available in summary form and are mentioned where applicable, however, the thrust of the history lies in recounting fighter operations, specifically P-38 operations ranging from the West Coast in 1942 to England, North Africa, Sicily and Italy during the years of WW II. During these years the 1st Fighter Group served additionally as a source of talent for both the 82nd and 14th Fighter Groups. Taken together these groups comprised the P-38 fighter units in the Mediterranean Theater of Operations. Early in 1942 the 1st provided senior staff, experienced pilots and NCO's for the newly formed 82nd. In May 1943 the 1st provided the experienced pilots needed to restaff and reactivate the 14th Fighter Group in North Africa.

ACKNOWLEDGMENTS

This venture into history began as a retirement project in 1987 commencing with a history of the 94th Fighter Squadron during the years of WW II. That in turn led to an interest in the history of the 1st Fighter Group. There isn't space to list the names of everyone who contributed to this task; most are given in the text or photo acknowledgments, however, I must mention and thank Jack Ilfrey who was instrumental in getting the Group history started. Jack made the connection between Phalanx and me and we went on from there. Jack also made available his own copious files and photos of events in the early years of the war both for the 94th and the Group.

The military events and actions described in this history were verified by a review of the three Squadron and the Group mission reports and records on file at the National Archives, Washington D.C. and the microfilm records obtained from the Albert Simpson Historical Research Center, USAF, (now the USAF Historical Research Center) Maxwell AFB. Some of the photographs were obtained from the National Air & Space Museum, Washington D.C.; the rest and, by far the majority, were contributed by members of the Group and other interested friends. Information on casualties was obtained, in part, from the American Battle Monuments Commission, Washington D.C.

Several people were particularly helpful with solutions to specific problems of identity and activity within the squadrons. Tom Maloney, Joel Owens and Francis Lawson helped with several questions about the 27th. Charlie McCann, Bucky Harris and P.E. MacAllister helped with the 71st and Stan Karpowicz of the 94th and his son Paul sent me more photos than I knew existed. Bob Vrilakas was instrumental in providing information both on the 94th and more general information. Clyde Gabler was particularly helpful on group matters with his What Did You Do In WW II Grandpa? Joel Owens and Vincent Rethman helped with the role played by members of the 1st in reactivating the 14th Fighter Group.

Since the 94th history had been written and I was a member of that squadron I felt a particular urgency to afford equal time, so to speak, to the Petdog and Cragmore constituencies, i.e., the 27th and 71st squadrons. I hope I was successful. I must also recognize those who contributed so much to the 94th history. Those contributions became a part and in fact formed the basis of this history.

Particular thanks are due my wife Phyllis who helped by doing a lot of copying, note taking and mailing. Her suggested retirement project assumed a size and complexity that neither of us expected when this venture into history commenced in 1987. Thanks also to Celia Dollarhide who helped with some identifications.

Finally I would like to thank all of you who sent me information about events of some fifty years ago. I hope I've used it to your satisfaction to present our history of those years. It's a history which I believe must be recorded to describe the events of a few short years when we literally, along with a good many others, determined the course of world history.

John Mullins

FOREWORD

This is a factual history of the 1st Fighter Group, a unit which had its origins in World War I. This history covers the entire period of World War II and sets forth the extremely adverse and variable flying and living conditions under which all of the members of the 1st Fighter Group lived, fought and many still managed to survive. It tells of the countless missions flown against a very capable and organized enemy, all of which were only made possible by the loyal, dedicated and unending support provided by Group personnel.

The history begins with the deployment of the three squadrons flying P-38's to California from their home base at Selfridge Field, Michigan. Subsequently the Group was deployed to England and successfully flew across the Atlantic. Later the Group completed a second, long, overwater flight from Lands End, England to Oran, in North Africa in support of Operation Torch.

Had it not been for the loyal and dedicated fighter pilots and the outstanding and dedicated gound support personnel, all of whom faced very challenging conditions on a daily basis, history might have taken a different, tragic course. The German Empire might have stretched from Lands End to the Aleutians and the Japanese Empire could have dominated all Asia as well as the Pacific Rim.

It is hoped that this recounting of the experiences of the flight and support personnel of this outstanding 1st Fighter Group during World War II will inspire future generations of Americans who want to serve their country.

Col. Ralph S. Garman, (USAF Ret.)

[Col. Garman joined the 1st Fighter Group in 1938 and served in various capacities in the 17th and 94th squadrons as well as group headquarters. He was commander of the 1st Fighter Group from December 1942 until September 1943.]

PROLOGUE

The past is indeed prologue to paraphrase Shakespeare. In this instance the past is long, illustrious and deserves mention. What is now the 1st Fighter Wing is an outgrowth of the 1st Fighter Group of WW II which in turn was an outgrowth of the 1st Pursuit Group of WW I. The original unit in France was the First Pursuit Organization and Training Center established near Paris in January 1918. In May 1918 this organization became the 1st Pursuit Group. The original organization melded into the second and ceased to exist.

While the 1st Pursuit Group was progressing from a training center to a pursuit organization in 1918 its component squadrons for the most part predated it. The 94th and 95th Aero Squadrons were formed in Texas in August 1917. Both assembled in France late in 1917 and commenced what might be termed operational training at Issoudon about 120 miles south of Paris. Both squadrons were assigned to the 1st Pursuit Organization in February 1918. The fledgling American pilots began to fly patrols each led by an experienced pilot. Somewhat disconcertingly the patrols were flown in planes with no armament, by pilots with no gunnery training.The 94th and 95th squadrons were soon joined by the 27th and 147th squadrons so that the 1st Pursuit Organization consisted of four squadrons by April of 1918. On Sunday April 14th 1918 the group could claim its first victories in the form of two German aircraft downed by Lieutenants Campbell and Winslow of the 94th Squadron. Obviously the armament situation had been remedied.

In May 1918 the 1st Pursuit Group was formed out of the 1st Pursuit and Training Organization. In June the entire unit, group headquarters and four squadrons, relocated to an advanced airfield close to Chateau Thierry and their baptism of fire really began. The German Aisne-Marne offensive was supported by heavy concentrations of aircraft including the famed Richthofen Flying Circus. The 1st Pursuit Group was a part of the 1st Air Brigade opposing the attack. Compounding the problems, about this time the Group began transitioning from the Nieuport 28's to Spad XIII's. Each plane had its adherents but overall the Spad was a heavier more powerful machine.

Late in the war, in October 1918 the 185th Night Pursuit Squadron was added to the 1st Pursuit Group. Night fighters were a real innovation. Their target was the German bombers that usually flew by night and constituted a general harassment behind the Allied lines. The night fighter tactic was simplicity itself. Climb to sufficient altitude, shut everything down, listen and glide. Enemy aircraft could be located (in theory) by a combination of engine noise and the sight of engine exhaust. If the plane lost too much altitude, restart the engine, climb back to altitude, shut down and repeat the process. The best the squadron could do was one probable but then it was in operation only about a month.

The 1st Pursuit Group ended the war with confirmed victories of 202 airplanes and balloons destroyed. By squadron the 94th had 67 1/2; the 27th 56; the 95th 47 1/2; and the 147th 31. This at the cost of 72 American casualties. At the signing of the armistice the group had participated in five major conflicts; Champagne-Marne, Aisne-Marne, St. Mihiel, Verdun and the Meuse-Argonne. The group was disbanded on December 24, 1918 at Colombey les Belles, France and its members ordered home for demobilization. On August 30, 1919 the 1st Pursuit Group was reactivated at Selfridge Field, Michigan. The Group at that time included the 94th, 95th, 27th and 147th Aero Squadrons. In the 1920's elements of the 103rd Aero Squadron, formerly the Lafayette Escadrille, were incorporated into the 94th squadron, the 147th was redesignated the 17th Aero Squadron and the 95th squadron would be permanently detached. These changes would reduce the 1st Pursuit Group to what would be considered the standard three squadrons. For at least the next decade the 1st would be the only pursuit organization in the fledgling air corps.

The 1920's proved to be a decade of training, consolidation and experimentation. The 1st Group served for some three years as the Advanced Pursuit Training School at Kelly Field and later Ellington Field in Texas.

Returning to Selfridge in 1922, the War Department sanctioned group and squadron emblems dating back to 1917-18. The group emblem was approved; the Great Snow Owl became the official emblem of the 17th; the Kicking Mule was approved for the 95th prior to its detachment and the Diving Falcon was OK'd for the 27th-a change from the original diving eagle, now a brand of beer. The 94th Squadron also changed it's emblem. The well known Hat-in-the-Ring had become too well known. It had appeared as a hood ornament on the short-lived Rickenbacker automobile and the Schwinn bicycle company secured a licensed trademark including the Hat-in-the-Ring. In view of all the commercialization the 94th gave up the emblem and adopted the Indian Head which had at one time been the insignia

This 1933 Selfridge Field photo shows the 1st Fighter Group arrayed by type: (l. to r.) Curtis P-6Es bearing the 17th Sq. insignia, Berliner-Joyce P-16s and P-12s. A new prototype Boeing XP-26 is at far left. (Warren Bodie)

of the Lafayette Escadrille. Since what had been the Escadrille was now a part of the 94th the new emblem was appropriate.

During the 1920's the group sponsored the Mitchell Trophy Race, took part in maneuvers, air shows, tactical demonstrations and what seemed to be annual cold weather operations, these latter usually in Canada or Alaska. In 1927 the 95th Squadron, one of the original units, was detached permanently and became, in time, a B-26 bomb group escorted by the 1st Fighter during North Africa operations in 1942-43.

Early in the 1930's, in addition to the depression, the 1st Pursuit faced one of its most challenging peacetime assignments - flying the U.S. mail. This occupied the group from February to June 1934. During this period the group lost two pilots and as many as ten aircraft. Overall the pilots were not trained for the type flying required; facilities, both ground and in the air, were primitive; and winter/spring weather could be daunting.

During this decade the group frequently lost both personnel and equipment in significant numbers as the fledgling air corps slowly expanded. Both men and planes went to the 20th Pursuit Group in 1932; to the 8th Pursuit Group in 1932 and this continued and accelerated toward the end of the decade and into the early 1940's. Cadres went to the 51st Group in 1941 the 80th Group in 1942 and the 82nd Group in April 1942, this latter even as the 1st was preparing to move to Europe. These moves were, in general, not helpful to morale.

Throughout the decade of the 1930's the 1st Pursuit became a proving ground for more and more sophisticated aircraft. The P-16 was introduced in 1934 just prior to the mail fiasco. By 1937 the group was flying the P-35 and 36 soon to be followed by the P-43 this latter the predecessor of the P-47 Thunderbolt. These machines were quasi-modern aircraft; all metal, enclosed cockpit, retractable landing gear. Some were used in the early days of the war at Pearl Harbor and the Philippines simply because nothing else was available.

In October 1940 the group underwent its final prewar organizational change. The 17th Squadron was detached from the 1st and sent to the Philippines where it was subsequently overwhelmed by the Japanese. In January 1941 the 71st Squadron was activated at Selfridge and assigned to the 1st Pursuit Group. While the 71st designation is 17 spelled backward there is no evidence that the War Department acted deliberately.

P-26B over Selfridge area, 1935. (Britzke)

A lineup of 17th Sqd. P-26's "Peashooters" at Selfridge Field 1936.

On the other hand there is no evidence that it did not. The latter makes a better story and imparts a degree of human interest to War Department decisions. Terminology also was undergoing change. In February 1940 the group was designated 1st Pursuit Group (Interceptor). In July 1941 it became the 1st Pursuit Group (Fighter) with the squadrons designated accordingly.

The new 71st Squadron was something different in the 1st Fighter Group. As Bob Naismith says, "It had no traditions, no Hat-in-the-Ring, no Pouncing Eagle and, in the beginning, no emblem at all." With only four exceptions, Harris, Wells, Miles and Eiland, the squadron was made up of Class 41-E graduates from Brooks Advanced, described by Naismith as the Brooks Brothers. The 71st was also isolated from the other two squadrons both in location and in aircraft. It was located at the extreme north of the flight line at Selfridge and equipped with P-43's and P-35's in contrast to the P-38's of the 27th and 94th. Since Grambo, the Squadron C.O. and Jenks were the only two old hands at flying either aircraft, the newcomers more or less checked themselves out and came early to realize that their futures as pilots depended heavily on the squadron's experienced ground crews.

By mid-year 1941 the 27th Squadron, 1st Pursuit Group had taken possession of the first of a new aircraft radically different in design and performance. This was the P-38 Lightning, a twin-engine, high performance aircraft with which the group would fight WW II. While the 27th received the first P-38's the new aircraft was eventually distributed among all three squadrons. Overall, however, the group, late in 1941, was equipped with a mix of aircraft including the new P-38's as well as P-43's, P-35's and some P-40's.

In firepower alone the P-38 represented a radical departure in military thinking. The P-38 armament included four 50 caliber machine guns and one 20 mm cannon, all clustered in the nose of the aircraft. At about this same time one 37 mm cannon was considered by some an adequate ground anti-tank weapon and well into the 1930's some fighter aircraft carried only a pair of 30 caliber machine guns or one 30 and one 50 caliber gun.

In the late summer and fall of 1941 the 1st took part in large scale Army maneuvers. The 27th squadron departed Selfridge in August for Beaumont, Texas, Savannah and Columbia, South Carolina. The 94th and 71st left for New Orleans, Savannah and Spartanburg with the 94th going to Portland,Maine in October, 1941. The three squadrons were back at Selfridge in early December after several months of training, although the 94th had been alerted (in November) that it would be assigned an additional 90 days detached duty on the west coast starting in December 1941.

Lieutenant Bob Naismith, at the time a new pilot in the new 71st Squadron, recalls the Louisiana-Carolina maneuvers and provides a participant's view of the state of readiness of the military, at least of the Army and Air Corps in the months just before the beginning of World War II. The following are gleanings from his journal and Form 5.

The 71st Squadron departed Selfridge in a haze of blue smoke. Naismith, with three and a half hours flying time in a P-35 and sixteen in a P-43 was assumed to be "combat ready". The squadron's first enemies were lying in wait as the twenty-four planes landed at Beaureguard Field, a concrete strip in a sea of mud near Alexandria, Louisiana. On landing the pilots were invaded by a) energy sapping, muggy heat and b) clouds of mosquitoes all with voracious appetites. Naismith's story continues:

"On the night before hostilities were to begin an Army staff officer wearing a blue arm band briefed his warriors. He informed them that the Army was putting a force of half a million men into the field divided into two elements; the First Army-the Red Force and the Third Army-the Blue Force. The mission of the Reds was to attack, that of the Blue to defend. The 71st were the good guys-on the side of the Blues to give them air cover.

"In the morning the squadron's eager beavers assembled and awaited orders. There were none. In the absence of directives from above our exec. directed his charges to fly around for a while, get acquainted with the territory and intercept any airplane coming anywhere near the base. As the week wore on it became clear that the first day's pattern would prove to be the norm for much of the rest of the games. Officially the squadron was a minor element in a giant maneuver. Actually, the squadron members, as individuals didn't have the slightest idea of what they were supposed to be doing. The Red Forces were somewhere out in never-never land but the mosquitoes were the in-place blood foes. It has been said that a good commander knows when to retreat. After a few nights of swatting and scratching Rudy (Rudell) admitted defeat. He arranged for all pilots, including himself, to spend nights in Army barracks on the airfield in New Orleans. Mornings, all got up, breakfasted and climbed into their airplanes for a boring hour and fifteen minute flight to Beaureguard Field where they refueled and waited for something to happen. At day's end they formed up and flew back to New Orleans. Between times the air was theirs. Ground to air communication was nil. There were no air controllers and, had there been any they couldn't have reached those who were airborne.

"There was one exception however. The squadron was assigned the mission of locating and destroying enemy tanks at a given map coordinate. Referees on the ground would rate the outcome. With no predetermined doctrine for ground attack our number two created his own on the spot. He would lead to the target, signal to form into a right echelon, circle twice in a left hand orbit, then waggle his wings. On cue each was to launch his own attack. They would simulate firing machine guns on the first pass, then pretend to drop bombs on the second.

"Six olive drab Army trucks with red markings on top and the word 'TANK' painted on their canvas sides made the job of locating easy. A ring of anti-aircraft weapons that looked for all the world like hand held stove pipes surrounded the vehicles. (It was later learned that they were indeed stove pipes with the word 'ANTI-AIRCRAFT' painted on them.) The aircraft ordinance was hardly more impressive. Only two of the airplanes had any guns at all and those were two WW I Browning .30 caliber machine guns synchronized to fire between propeller sweeps. Of bombs or bomb racks there were none.

"On signal the squadron peeled off to work their quarry over. In their enthusiasm each went low enough on both runs to cause defenders to run for cover in the ditches. This was the 71st's war and they were expected to take it seriously. Back at Beaureguard the victors made hotshot peel offs for spaced out landings just like in 'Wings'. Reporting noisily for debriefing their intelligence officer was hard put to calm them down. Having determined the number of vehicles observed he succeeded in narrowing claims from twelve tanks to six. His final question was 'Where's Lieutenant Strozier? He had been the last man in echelon as the squadron broke for landing. And he was no longer present."

Lieutenant Strozier was in an inverted P-43 crashed at the end of the runway. Luckily he was unhurt but his aircraft was a total washout. Not only that, the war games umpire saw the whole picture quite differently. He ruled that every single airplane had been blasted out of the sky before any of the 71st pilots had fired a shot. The tank busting caper was the only mission the squadron was given during their month long stay in Louisiana. Drawing a parallel Bob Naismith goes on to include the intelligence that on September 16, 1941 in the land of the rising sun, the Japanese were running a board game to test their overall aggressive plan. Their Red Forces won decisively over their simulated Blue forces defending Hawaii.

For the 71st Squadron it was on to Savannah, Georgia for more training. Savannah Army Air Base was, unfortunately, not quite prepared for the 71st. Grading equipment was stirring up dust clouds around a dismal looking tent city as the squadron arrived.

The first mission of the 71st, as explained to them by, in Bob Naismith's terms, "an Army liaison officer, a paddlefoot of the first water." He described an electronic device (probably a primitive radar) which would direct searchlights onto high flying aircraft, in this case a B-17. Three 71st Squadron flights would be flying area coverage during the night. Whichever flight was closest to the illuminated B-17 would close in for the kill. The

Sgt. Fred Quick in 17th Sqd. P-35, Selfridge late 1930's. (Quick via Warren Bodie)

71st pilots were qualified for night flying by virtue of floodlight assisted takeoffs and landings at Selfridge. Savannah had no lights, the runway was flint gravel, P-43's had no landing lights and very poor radio capability. Nonetheless gasoline soaked, sand filled buckets would be placed along the runway for illumination on takeoffs. After several tries nothing was accomplished. The B-17 failed to show, next it did show but the electronics failed and so it went. The days were given over to practice cross-countries with the single exception of one high altitude experiment.

Lieutenants Naismith and Morefield were informed that they had volunteered to perform a high altitude test to check the electrical system in the P-43. They hadn't volunteered of course but it sounded better. Oxygen equipment in the P-43 consisted of a rubberized rebreathing bag strapped to the pilot's helmet. Oxygen was fed through a rubber tube from a high pressure bottle and metered through a manual valve that had to be adjusted in accordance with altitude changes. Neither pilot had ever flown at heights which required oxygen.

The mission went about as expected. Morefield's ship performed perfectly. He reached 37,000 feet and returned uneventfully. Naismith wasn't as fortunate. After some instability his electric system went haywire at 37,000 feet producing a shortcircuit, blue smoke and then black smoke but luckily no fire. A blown cylinder produced most of the smoke and also a dead stick landing from 37,000 feet with an oil covered windscreen and, of course, no power. That completed the stay at Savannah. The cost - three more damaged airplanes and no combat training.

The 71st headed for Spartanburg, South Carolina. In Naismith's Japanese parallel, on November 5th the "Japanese held a successful dress rehearsal for their attack on Pearl Harbor. It was A-day minus 32.

"Spartanberg was supposed to be air-to-air training but the only planes in sight were the 71st own P-43's. Pilots took turns sitting in cockpits of "ready" planes on alert status. At least once a day the tower would fire a Very pistol as a signal to crank up and take off. There were never any attackers to defend against but the brief scurry of getting off the runway was a welcome exercise in alleviating tedium.

"The dates of November 11 through 25 were set aside for combat in the heavens. Missed connections and a spate of bad weather wiped out those plans. Finally the operations tent burned on November 20 taking with it all the squadron records. For the 71st the war was over. The only other incident of note was a visit by a British veteran of the Battle of Britain who managed to get lost on a cross country, ran out of gas and as he put it, 'pranged the bloody kite'."

The Fall maneuvers wound down in late November and the 71st headed back for Selfridge. "November 30 was moving day again and heads were filled with plans for a relaxing Christmas season. They were seven airplanes poorer, pilots had logged over 200 hours apiece and engines were beginning to show the effects of sucking up flint dust. The hard working ground crews had proven that they were masters of the art of maintaining equipment under the most difficult of field conditions.

"Across the Pacific it was A-Day minus 7. Japanese training maneuvers were over. They had different plans; their forces were on the high seas."

In early December 1941 as the Japanese battle fleet was steaming toward Hawaii, the 1st Fighter Group headquarters and two squadrons were recently returned to Selfridge. One squadron, the 94th, was en route to the west coast as an extension or continuation of maneuvers and not in anticipation of the events of December 7th at Pearl Harbor. The stage was now set.

P-43's of the1st Fighter Group at New Orleans in September 1941 at start of Carolina Maneuvers. (Warren Bodie)

1

THE BATTLE OF CALIFORNIA

December 7, 1941

On this early winter Sunday the members of the 27th and 71st Squadrons were still getting reacquainted with their permanent post. Both squadrons had, only a few days ago, returned from several months of maneuver activities in the south and southeast. The 94th, in contrast, was landing that afternoon in El Paso on its way to ninety days detached duty at March Field in California. The 94th had been advised about this assignment in November while still on maneuvers so, in one sense, the West Coast was an extension of maneuvers with only a brief stop at Selfridge.

The 94th squadron was informed about the Japanese attack on Pearl Harbor on landing in El Paso. After a restless night the squadron (ten P-38s and ten P-38Ds), led by First Lieutenant Ralph Garman, took off the next morning for March Field stopping in Tucson for gas. As the squadron approached California it was vectored south to San Pedro, landed, then redirected to North Island finally down at 1635 hours. The ground contingent of the 94th was also en route, some by air and some by train.

As soon as the reality of Pearl Harbor became known the 27th and 71st and the remainder of the 94th Squadron at Selfridge received orders directing them immediately to the West Coast. The air echelons of both squadrons left after some frantic preparations and flew to the West Coast landing at Cheyenne, Wyoming for gas. The two squadrons were flying a mix of airplanes, some Lightnings, some P-35s and P-43s, the latter a woefully underarmed forerunner of the P-47 Thunderbolt.

Departure of the ground contingents remaining at Selfridge was a hectic affair, doubly so given the recent return from maneuvers. After the planes remaining at Selfridge had been prepared and the air echelons departed, all equipment had to be crated and loaded and, of course, all personal belongings washed, cleaned and packed. In the 94th, Ray Ainsley's wife Sylvia tried to hurry things along by drying some of Ray's freshly washed clothes in the oven. Unfortunately they overdried and came out nicely singed. No clothes dryers in those days! A total of 160 enlisted men and six officers departed Selfridge by train on December 9, 1941. Clyde Gabler gives us a hint of the scope of this operation. He notes, "The first section of the ground echelon left Selfridge after working 24 hours straight to load and equip a mixed railroad train of 28 passenger and baggage cars plus several flat cars... all under the command of Capt. "Bull" Durham." (As an aside, Bull Durham was a brand of roll your own cigarette tobacco). Minutes of the 71st squadron note a 6:30 PM departure of 144 men, 2 men attached, one flying cadet and one dog-a mascot named C.Q. C.Q. unfortunately jumped ship in New Mexico but arrangements were made to ship him on if and when located.

By December 17th all personnel of the 1st Fighter Group were together at North Island. This latter was beginning to resemble an aircraft carrier with planes of every description, Army, and Navy coming and going.

During December the individual squadrons flew patrol after patrol and found nothing. Early in the month a Japanese attack of some sort was expected almost daily by a very jittery populace and an untried military. As patrols continued the 94th lost Lieutenants Foy Robertson and William Northern on December 21, 1941. The two collided attempting a cross-over and went into the ocean. Not long after this, on December 28th, Lieutenant Allen Scales, a pilot in the 71st squadron disappeared over the Pacific. He had been on a routine patrol and machine gun practice flight.

Christmas came, patrols continued and some measure of confidence in the future began to reappear. However, Christmas of 1941 was scarcely a happy affair although a traditional dinner was served and each man received a package containing fruit, nuts, candy and cigarettes. While the initial Japanese invasion scare had begun to subside, there were no family members present, security was tight and the days were grim indeed. However, unit members made the best of it. Several of them (according to Ray Ainsley of the 94th) went into San Diego and returned with a tree, decorations, four gallons of wine and glasses. All of this was taken past the gate guard under the guise of radios to be sent home. Seems strange, but subsequent stories to various future guards would be equally strange.

Christmas passed and the new year of 1942 arrived. What would it bring to the 1st? What indeed would it bring to the United States? The news was grim on all fronts. The Pacific was a disaster area. In the opposite direction, England was hanging on by two threads called Hurricanes and Spitfires.

Given the crowded conditions at North Island, the need for some dispersal of aircraft and personnel became more and more necessary. As December ended the 27th, commanded by Major William Covington, moved to Mines Field in Los Angeles. Early in January the 94th, then commanded by Lieutenant Glenn Hubbard, moved to Long Beach Municipal Airport. Major Garman, with the Headquarters Squadron, was assigned to the Los Angeles Information Center and the 71st, commanded by Lieutenant Raymond Rudell, moved to the Grand Central Air Terminal at Glendale on February 12th, 1942.

Glenn Hubbard, the 94th Squadron commander, had an interesting and invaluable background. He went to England in May 1941 as an observer with the embassy assigned to the RAF to learn what he could. He flew both Spitfires and Hurricanes, talked with the combat experienced RAF pilots and studied their tactics. The RAF flew an off-center V formation with three flights flying abreast. In the early days of the U.S. entry into the war the 94th, and then the group, used the off-center V formation to replace the trail formation normally used by our pilots. Hubbard's point was that the #6 man or tail-end Charlie in a trail formation was generally a loser in any maneuver. The off-center V in time became the "finger four" formation used throughout much of the war.

The records indicate that Glenn Hubbard was some three weeks late reporting to the squadron on the West Coast and therein lies a tale. At the conclusion of the South Carolina maneuvers Hubbard was hospitalized about a week and discharged on December 7, 1941. Phoning Selfridge he was told to wait for a pickup. After some considerable delay he was retrieved by a B-18 and returned to Selfridge about 4:00 AM. Next day he went to headquarters to see about transportation to the West Coast to rejoin the squadron. He was informed by HQ that everyone else in the group had left in a rush, the barracks were in a mess and it was up to him to get the place cleaned up. With some help he got things cleaned up and crated and then went looking for transport to San Diego.

He found two P-38s, one stripped of everything even some of the cowling; the second one had a faulty right engine. Glenn then recalled that there was a P-43 on exhibit on the 5th or 6th floor of a building in downtown Detroit. He phoned Republic in New York to get some help, went in to town, got the wings off the P-43, took the aircraft sections down a freight elevator and headed back to Selfridge. On the way the driver stopped too quickly at a traffic light, the P-43 came forward into the truck cab and smashed the tail. Continuing to Selfridge, he found a crashed P-43 in the junk yard with an intact tail switched tails and, voila, one flyable P-43. But now two additional pilots from the other squadrons had straggled in. They located a second non-flyable P-43 but the Republic mechanic got it in flying condition. Now they had two flyable P-43s and Glenn Hubbard's P-38 with the bad right engine that no one could fix. Glenn finally got it off the ground, light weight, on single engine and the three of them flew to Wright-Patterson. So far so good.

The ailing P-38 right engine was diagnosed as a bad fuel pump and, of course, the people at Wright said they had no spare parts. Walking around the field Glenn found a hangar with several planes in it including one P-38. The mechanic said he had no authority to remove a fuel pump so the two of them went back to the P-38, removed the pump and installed it in Glenn's ailing right engine-Glenn to take the blame if some higher authority objected to the transfer. The Lightning was then towed out of the hangar ready to go. Unfortunately the guy driving the tow truck turned left too soon and put the left wingtip of the P-38 through the operations office window. Scratch one wingtip. Back to the P-38 source of the fuel pump, take off the left wingtip and install it on Hubbard's ship. Finally all three fighters took off and barely made it to Barksdale. One of the P-43s had a runaway prop. Called Republic again, they sent another mechanic, prop fixed. That completed, all three took off and actually made it to San Diego about three weeks late leaving a trail of scavenged airplanes behind them.

These early months of 1942 were occupied with an unending series of patrols which were uneventful but cost three pilots, all from the 94th during March and April. Lieutenant James Lauck was lost on an infrequent night patrol, Lieutenant Lawrence Scholtzhauer went into the ocean with runaway props and Lieutenant Walter Lichtenberger was killed in a crash landing on Catalina Island.

The pilots were on call four hours at a stretch but sometimes with three or four days without an alert people got lax. To combat this Hubbard as CO of the 94th, requested an alert at least once in every 24 hour period. Pilots were alerted by a bell system and, at the signal, they ran out jumped in their planes with parachutes already in place and took off. One day at chow a C-47 landed and pulled up to the hangar where everyone was eating. Thinking it was a supply delivery Hubbard walked out then noticed that the C-47 had curtains - no supply plane this. At that point out stepped General Kenney who said, "I'm the Japanese! Get your planes in the air!" The alarm bell rang, the pilots poured out of the hangar, jumped in the P-38's and were off in minutes. The General thought it was the best performance he had seen on the West Coast. He had been visiting bases using the stop watch technique to see what sort of response times would result.

The 82nd Fighter Group was being organized during this time and the 1st Pursuit Group was levied for officers and men to staff the new group. The 82nd Group would turn out to be our neighbors during much of the war, however this nearly incessant transfer of large numbers of men was not helpful to morale. In April 1942 even as the BOLERO[1] time was approaching the 1st lost its group commander Lieutenant Colonel Israel, twenty-three officers and 338 airmen to the newly formed 82nd Group. As an example of the impact of these reassignments the 71st squadron lost three Master, six Tech and seven Staff/Sergeants, experienced men, not easily replaced. Major John Zahn took over as group commander and Major Garman assumed the group Executive spot. In the final analysis the 82nd was almost completely staffed by men from the 1st Group from the group commander to flight leaders and ranking enlisted men. Aside from the flight leaders most of the 82nd pilots were staff sergeants an anomoly of the early Army Air Corps.

By late February coast inhabitants were beginning to calm down a bit and not see Japanese behind every palm tree. Not so on the morning of February 25th. Headlines on the order of:

"Submarine Shells Southland"

"Submarine Shells Near Santa Barbara Early In The Night"

Why would a Japanese sub lob a shell or two somewhere near Santa Barbara? It was eventually determined that the two shells had done no damage. Why would the West Coast panic anew and embark on a retaliatory plan that would uproot from their homes everyone with Japanese ancestry.

Bob Naismith, 71st Squadron, tells us how at least one minor event led to a host of dire consequences:

> "Soon after the Pearl Harbor attack, as a partial bulwark against this doleful scenario (a Japanese attack) the Army assigned its 1st Pursuit Group the responsibility of patrolling Southern California's coastline and serving as first line of defense against what popular newspapers ballyhooed as a certain attack. Around the clock aerial patrols were backed by air raid wardens who were to make visual observations of aircraft approaching the coast and batteries of anti-aircraft artillery to dispatch any and all aircraft deemed hostile, a scenario which caused the sky over Southern California to come unglued.
>
> "On the night of February 25th Lieutenant William "Big Bill" Newman of the famous 71st Squadron was on routine duty, picketed off Santa Monica. He was leisurely boring his way back and forth at an assigned altitude of 4000 feet, as he expressed it later, 'fat, dumb and happy,' when his electrical system failed and cockpit instrument lighting died out. Tootling around in the dark without gauges to guide him was no fun, so after giving the matter some consideration Big Bill fired up his red-banded signalling flashlight and took a reading on his instrument panel. He reported later that what he saw made his lower intestines clog his throat. His ammeter was registering full discharge. Not a happy turn of events. He knew that when the plane's batteries discharged both electrically controlled props would feather on him and turn his heretofore dependable flying machine into a non-flying chunk of machinery. That was no place for "Little Willie" as he called himself, to hang around. To make matters worse, his flashlight switch jammed in the 'ON' position and resisted all efforts to turn it off. During debriefing he admitted that in frustration he had cranked his side window down and heaved the offending piece of junk over the side. Bill then opened his throttles and departed the scene.

Jack Ilfrey, Long Beach, CA early 1942. Jacket with 94th Indian Head emblem. (Jack Ilfrey)

> "In the meantime a dedicated, sharp-eyed air raid warden on the coastline below spotted that tumbling red and white object and came to an instant conclusion. The Japanese had attacked the night before; they were at it again. He ran to his phone and called the Los Angeles Intercept Board to report that aerial combat was taking place over Santa Monica Bay with one plane shot down in flames. Intercept called Big Bill and got no response-his radio like his light system, was out. Aerial combat? One plane shot down? No response from Newman? A duty officer at Intercept took these items at face value and immediately invoked a red alert. In effect, gentle Big Bill had innocently attacked his homeland.
>
> "Given time the whole situation would probably have been sorted out and the alert cancelled. However, a few minutes later some poor Navy pilot came chugging in from Pearl in a PBY. He missed his identification zone north of Coronado Island and hit Santa Monica right on the nose. An unnamed AA gunner on red alert saw the juicy target and let fly, whereupon the startled Navy fly-boy had sense to douse his running lights. Ground marksmen then zeroed in on Mars, Venus, each others' shell bursts and the sky in general. Soon guns were firing from Long Beach to Malibu with only an empty sky to shoot at. As soon as the firing started someone in the chain of command ordered a Southern California blackout.

Master power switches were pulled hither, thither and yon.

"The Navy pilot did what any sensible young man would do. He took a reading on his location and opted to use his remaining fuel to head for a safe haven at the Salton Sea. Big Bill decided that trying to land on the by then blacked out landing strip in Glendale without lights or instruments and with guns going off all round would not be the wisest of procedures so he headed for wide open Muroc Army Air Base.

"The comedy of errors continued as phones began ringing in the apartment complex where three married officers were quartered relaxing in wedded bliss. It was the Intercept Officer calling. All standby off base pilots were to report immediately and prepare for takeoff. The intrepid three piled in to one trusty '41 Chevy and headed for the field. Within a few blocks Glendale police pulled the speeding vehicle over and asked sarcastically if the driver didn't know that there was an air raid on, that no private automobiles were allowed on the roads during air raids. Military ID's and flight suits didn't daunt Glendale's finest. They had their instructions and that was it. Shonie, always the great communicator, informed them in his best military voice that the pilots orders came from General Israel (a slight exaggeration, Israel had only just been made a light colonel) and demanded to know on what authority they were countermanding his orders. In the same tone of voice he told the driver to drive on. The police were left speechless.

"Another obstacle awaited at the field. A National Guard outfit had bivouacked on the inside of the perimeter fence while awaiting transhipment to the Pacific. When the night's fireworks had begun popping their CO had taken the initiative; he posted guards at strategic locations with orders that they were to protect the landing field, allowing no one through the gate. One of his grim faced citizen soldiers, holding a fixed bayonet, held the three at bay. He announced in a post teen-age squeaky voice that he had his orders. No one was to pass through their gate and only their CO, the Officer of the Day or the Sergeant of the Guard could change those orders. None of the three saw any point in pressing the issue against a bayonet on the end of a rifle held in shaky hands."

Eventually the alert was called off, flying was cancelled until daybreak at which time all available pilots were to scour segments of the Pacific to see if they could discover from whence the mysterious aircraft had come. While the pilots were thus engaged, enterprising citizens tackled their runway, paint and brushes in hand to camouflage the concrete. This, of course, put the runway out of commission and forced the pilots to land on a weed patch beside the landing strip. One P-38 unfortunately had a close encounter with, of all things, a foxhole dug the night before. A final tally of the night's work: five people dead, considerable damage from falling shrapnel, tons of lost food from power outages, frayed nerves and the unreported loss of one P-38. (The Accident Board charged pilot Naismith with 100% pilot error on the grounds that all the other pilots had missed the foxhole.)

Unfortunately the matter did not end there. Headlines continued to make totally ridiculous statements as did several officials who should have known better. For instance:

"Stimson Says 15 Planes over City"
"Secretary Of War Declares Craft Flown By Enemy Agents Over Los Angeles Area"

The February 25th incident was quickly escalated into a full fledged outcry against so-called "Jap" spies. The Japs had attacked. They were aided and abetted by local Japanese spies. Therefore all Japs had to be removed from Southern California as rapidly as possible. This was accomplished by Executive Order # 9066 which was implemented as soon as possible and citizens of Japanese ancestry were shipped out by the busload. Bob Naismith concludes his recollections by remarking, "In memory of Big Bill Newman's flashlight, had Pogo been around he might have at least whispered, 'We have met the enemy and he is us.'" Thus was the state of mind, both civilian and military, on the West Coast early in 1942.

In April Eddie Rickenbacker[3] paid a visit to the squadrons of the 1st. Eddie witnessed a flight take off for patrol and said a few words, the whole business recorded and shown on Pathe News, good for the home front. In part, as a result of his visit, a move began to have the Hat-in-the-Ring (World War I unit emblem)reinstated as the official insignia for the 94th squadron. The approval of this took place later in the year. Rickenbacker was on a tour, at the request of General Hap Arnold, to see how things were going and they were not going well. Rickenbacker apparently returned with a litany of complaints about curtailed flying time, few spare parts, poor distribution of supplies, i.e., a P-38 would be out of commission for weeks due to a lack of spare parts. There was nothing wrong with the pilots' morale that couldn't be cured by more flying time.

On May 17, 1942 the departure from California began. The air echelon was off for Dow Field near Bangor Maine, first being outfitted with new P-38Fs at Lockheed. The ground echelon left by train for Fort Dix, New Jersey. The departure of the entire 1st Fighter Group occupied three days, with all hands away by May 19th. The air echelon consisted of eighty-five P-38Fs and pilots, sixty-five crew chiefs, an armament officer, weapons specialists, communications and instrument specialists. The ground echelon included ninety-nine officers and 872 enlisted men.

The cross country train trip was uneventful and not unpleasant with prices for snacks and soft drinks at ten

L - R Eddie Rickenbacker, Sgt. George Gemeiner, Lt. J.C.H. 'Bugs' Lentz. (George Gemeiner)

Rickenbacker autographed Hat-in-the-Ring displayed by Capt. Hubbard, 94th C.O., T/Sgt. Spitler. (Lenhart via Ilfrey)

to fifteen cents. These prices were established and enforced by a federal agency, the Office of Price Administration, known for short as the OPA. The objective here was to prevent overcharging of GI's in transit.

In this sort of division the air echelon included the planes and pilots together with a core group of mechanics, armament and radio specialists with their equipment. The ground echelon included, of course, everyone else; administration and supply personnel, the remainder of the mechanics, radio people and so on. In theory this division permitted the air echelon, travelling in P-38s and (usually) C-47s to operate independently for a week or two until reunited with the ground contingent. In practice the two week period was almost always exceeded. That was certainly true in May 1942.

Rickenbaker lecturing pilots on tactics, Long Beach, CA. (Lenhart via Ilfrey)

Eddie Rickenbacker presenting signed Hat-in-the-Ring emblem to 94th at Long Beach, April 1942. (Lenhart via Ilfrey)

The ground echelon departed the United States via the Port of New York, leaving on the *Queen Elizabeth* on June 4, 1942. The QE, because of her speed, sailed unescorted across the Atlantic. Several officers and men were assigned as blackout patrol which was a very serious matter on an unescorted troop ship even with the QE's speed. Dewey Prince, a member of the 71st ground contingent describes a little of the trip:

> "My group was assigned a first class cabin on A deck. It was a cabin intended for two people. There were sixteen of us in the room in four sets of bunks four high. Since each man had a barracks bag it was so crowded that we used the bathroom to store our barracks bags in. As a young man from a small town in Texas I had never been on a boat larger than a small ferry boat. The Queen Elizabeth was big. When we looked for the mess hall we found a line so long we couldn't believe it [two meals a day]. We spent most days standing in line for meals or in the PX line. We also had several boat drills."

Five days after departure the Queen docked at Gourock, Scotland outside Glasgow. The 1st Fighter Group, in part at least, was in the British Isles. The rest of the outfit was a long, long way behind them. The west to east route of the P-38s was: El Paso, Texas, Barksdale, Louisiana, Patterson Field, Ohio, Mitchell Field, New York to Bangor, Maine.

The air echelon in the meantime was caught in an "are we or are we not" situation at Dow Field in Maine. In the Pacific the Battle of Midway had begun and the Aleutians were invaded. The War Department thought it prudent to bolster Pacific Coast defenses. On June 5th the 94th, the 27th and the 71st squadrons departed Dow for Morris Field in North Carolina; the first leg of a return trip to the West Coast despite the fact that everyone else was on a fast ship heading east! That evening the danger in the Pacific was past at least for the time being, and new orders were received the following day returning all three squadrons to Bangor.

Since the 1st Fighter Group[2] was somewhat crowded at Dow Field, part of the unit moved to a field near Houlton, Maine. Houlton is some 115 miles north by northeast of Bangor. Much of the time at Houlton and at Dow was spent in practice formation flying with a B-17. Speed averaged about 160 mph.

Flying heavy bombers across the Atlantic was not considered much of a problem but fighters were something else again. With their limited range and meager navigational equipment fighter aircraft were not considered candidates for transatlantic flight. The P-38 was the exception. Twin engines and the newly developed external fuel tanks gave the Lightning the "legs" and the safety margin for transatlantic flight employing B-17s as navigational guides. The B-17 would lead the way with the P-38s flying formation on each wing. That, of course, was the theory since such a flight had never been carried out. Also theoretical were slow climbing speeds, icing, ice crystal fogs and a few other far north weather phenomena. Calculations used for flight planning assumed that the P-38 could fly for six hours at 200 knots at 15,000 feet. Starting with 570 gallons of fuel there should be an extra hour of flying time in reserve.

1. Cross Atlantic flights.
2. The old pursuit designation was changed to fighter on May 15, 1942
3. Leading American ace of WW I.

2

MISSION BOLERO

The 1st Fighter Group, along with sister P-38 units, the 14th and 82nd Groups, were part of the initial buildup of European based U.S. air forces, dubbed Operation BOLERO. The planned route across the Atlantic began at Dow Field thence to Goose Bay in Labrador. From Goose Bay the route went to Bluie West 1 or, less frequently, Bluie West 8 in Greenland, then to Reykjavik, Iceland and finally to Scotland. [1]

Transition to Goose Bay began in late June for the 1st Fighter Group air echelon with an intermediate stop in Presque Isle, Maine. The 27th led the way departing Bangor on June 27th. The 94th flew from Presque Isle to Goose Bay on July 4, 1942. Flying time for the trip was three hours, five minutes according to Jack Ilfrey's log . Goose Bay in the summer of 1942 left a great deal to be desired. The field was still under construction, manned by a few Canadian soldiers with an additional few Indians and Eskimos as neighbors. Food, what there was of it, was not good. The only thing present in any abundance turned out to be huge mosquitoes.

On July 2nd the group began the leg to Bluie West 1 in Greenland. Headquarters flight took off followed by the individual squadrons. The 27th made an uneventful crossing on July 2nd and the 94th departed Goose Bay on July 6th. While most of these flights were uneventful, one became an adventure.

The 94th Tomcat Black flight landed at BW-1[1] after a flying time of 4:15 according to Jack Ilfrey's log. Bob Wilson leading Tomcat Green flight was not as fortunate. Green flight included, in addition to Wilson, Lieutenants Carl Rudder, Brad McManus and J.C. Harrison Lentz. Tomcat Green together with their B-17 left Goose Bay on the morning of July 6th immediately behind Tomcat Yellow led by Dallas Webb also with their B-17. Climbing out over the Davis Straits they gradually transitioned from a view of blue water and icebergs to low lying overcast. This was not particularly disturbing because of the lead B-17 with its navigator.

The cloud layer thickened and slanted imperceptibly upward until, after an hour's flying Tomcat Green had reached an altitude of 12,000 feet, then it was 18,000 feet and getting cold . They next encountered a wall of clouds extending to at least 30,000 feet. Both flights turned north. Flying through ice crystal haze the B-17's were leaving wakes like miniature snowstorms. At 20,000 feet and 160 mph the turbo's kept cutting in and out making formation flying like operating a hand pump, according to Harry Smith. Forty minutes later and 100 miles off course the cloud barrier was still there. They could not go around it nor could they obtain any weather from Bluie West-1.

Brad McManus described the P-38s flying formation on the B-17 as, "hugging that Mack truck like a flock of lost chickens."

Both flights then began to reconnoiter lower altitudes. The B-17 from Tomcat Yellow reported bottom at 10,000 feet however this was only one of several cloud layers. The two flights became separated at this point with Tomcat Black headed first west toward land then back to Goose Bay culminating in a tough eight hour flight only to arrive back at their starting point. And what of Tomcat Yellow? They finally landed at Bluie West 8 some 400 miles north of their destination of Bluie West 1. Nonetheless they were in Greenland and they were on the ground.

Overall the group was badly strung out. Colonel Stone and the headquarters flight reached Scotland on July 9th and part of the 27th reached Iceland on July 6th.

After six more days of bad weather Tomcat Green and White flights flew uneventfully to Bluie West 1. The 94th was now all in Greenland albeit at two different locations some 400 miles apart, the 71st was at BW-1 and the 27th was in Iceland.

Bluie West 1 was an airfield consisting of a steel mat runway set in an amphitheater of ice cap rising to

L - R Ilfrey, Widen, Humphrey on ice cap above Bluie West 1 Greenland, July 1942. The headgear is mosquito netting. (Jack Ilfrey)

In flight over Greenland before the forced landing, one P-38 flies close formation on a B-17 "navigator". (J. Bradley McManus)

inland mountains. The field was actually some 40 miles inland on a fjord called the Tunundgliarfik Fjord. The general area was identified by the equally jawbreaking name of Narsarssuak. The runway was approached over water and ran uphill. The end of the runway was about 110 feet higher than the fjord end. Take off was the reverse - downhill and out over the water and circle to gain altitude. Six P-38s and one B-17 made it to Bluie West 1 on the first try. One flight reached BW-8 and two more were grounded in Goose Bay for nearly another week. Such were the difficulties of flying in the far north in the spring of 1942.

For those already in Greenland the next few days were occupied by fishing, hiking about the icecap - a pretty fatiguing experience - and a brief visit to a nearby eskimo village. A visit that was neither enjoyed nor repeated.

The 94th departed BW-1 on July 14th headed for Reykjavik, Iceland (code named Indigo). Tomcat flights Black and White proceeded directly in company with their B-17s. Green flight left BW-1 on the same day and headed for BW-8 to join Tomcat Yellow. The two flights would then depart together for Iceland leaving BW-8 on July 15th at 3:45 AM. The Greenland to Iceland leg of the BOLERO Mission typified the best and the worst of North Atlantic sub-arctic weather and flying. The BW-1 contingent made the trip in 6 hours under almost ideal conditions. For the flights departing Bluie West 8 one day later, the exact opposite was true.

Tomcat flights Green and Yellow left BW-8 (less Lentz and Starbuck) and crossed Greenland uneventfully. As they continued out over the Atlantic the weather grew progressively worse. With icing conditions and solid overcast they reversed course to return to Greenland. Once back above the icecap they again faced solid overcast. BW-8 reported as closed. BW-1 reported open, however BW-1 was at a distance nearly equal to their remaining gas supply - about two hours flying time. Bob Wilson's post event recollections (somewhat abbreviated) describe the final act.

"We cruised at 18,000 feet for an hour and, at 8:15 reached the eastern coast of Greenland once more. At that point we had to fly over the top of the overcast which lay on the surface of the icecap. At 9:00 AM, believing ourselves on course, we called for weather in plain language from the ground station at Bluie West 8. We were in an emergency and needed to know the weather at that station. Between us and BW-8 lay a solid front of ice and snow clouds. At 9:45 we received a report that BW-8 was closed in. We then called for the weather at BW-1. This came through with the report that the weather was clear and unlimited. We were beginning to worry at this point about our gasoline supply. We knew we had a couple of hours left. Our ETA for Bluie West 1 was 12:00 noon; the time was 10:00 AM. It would be close.

"Until the time of the clear report from BW-1, we had been circling over the ice cap at a clear spot where we could see the ice. We were discussing among ourselves whether we could set the ships down on the ice with wheels down. We would try flying over the ice and touching the wheels to the surface. If the surface felt solid we would cut the engines and land. If not we would all land wheels up.

"We headed on a course for BW-1 and flew for two hours at which time we should have seen mountains. We did but they (the mountains) were on our left but they should have been on our right since we were headed on a southerly course and BW-1 was on the west coast. We had already dropped our auxiliary gas tanks to reduce drag. The mountains on our left warranted our investigation because they were the only contact with the ground for the past two hours. When we arrived at the mountains about 20 minutes later we found that there was one clear spot and through it we could see the shore of Greenland. We were sure it was the eastern shore and through some freakish event we had gotten off course and hit the wrong shore. Brad (McManus) called and said he had only 20 gallons of gas left and was going to set down on the cap close to the coast while he could still see the cap. We has also decided that we would all set down at the same place when we had to set down at all. This would facilitate rescue as well as furnish mutual support and shelter in the bombers.

"Brad pulled out from the formation and I went with him. Rud joined Spider's (Webb) flight to see if they could discover where they were. We went

Lieutenant Brad McManus (left) and his P-38, the only one to flip upside down on landing and Lieutenant Carl Rudder. (J. Bradley McManus)

lower and lower and still could not judge the altitude of the ice cap. It looked packed and hard immediately under us. I was at 3,000 feet at this time and judged myself to be about 100 feet from the surface of the cap. Brad was lower than I was. He kept circling lower and lower and found that he was also unable to judge his distance from the surface and was feeling his way in. He put his wheels down and dragged his ship over the ice feeling it out for a landing. He rolled along the surface for some 75 yards and was making a beautiful landing when all of a sudden his nose wheel broke off, he flipped upside down and stopped in his tracks. There was a puff of smoke and then no movement. I was flying immediately beside him at the time and the first thought through my head was that he had been killed instantly. I leveled off and circled about 10 feet off the ice to see if there was any sign of fire. R.H.(Wilson) had been behind me and had landed at the same time Brad flipped over. He was half a mile away. I remember calling Joe and telling him to get the other flight back and land them. Spider(Webb) came back and ordered all pursuit ships on the ground (ice) and the bombers to circle and send SOS and our position. We had a shot at the sun and had the exact position at that time: 20 degrees N. Latitude, 20 degrees W. Longitude I circled Brad's (McManus) ship again and saw R. H. on his way to it. On the next pass I saw someone by the tail of Brad's ship and R.H. still on his way. I circled and landed at the same time as Rud and Spider. I landed wheels up as all the rest did and skidded to a halt about 200 feet from my first contact with ice and soft snow. I was completely covered by the flying snow powdered up by the props.

"The distances were very deceiving in this land of ice and snow. Because of this we had to walk about twenty minutes to reach Brad's plane.We found him standing up and apparently in good shape with no visible cuts or bruises. We were amazed. He seemed badly shaken up however. We asked him if he was alright and he answered that he thought he had cut his arm a bit. We stripped his shirt from him and found a somewhat deep gash in his right arm in the fleshy part of the muscle. I had brought my first aid kit from

One of the B-17 Flying Fortresses that accompanied the P-38s in the mass forced landing, "Big Stoop" became a survival hut. This one was dubbed "Hotel Big Stoop". (J. Bradley McManus)

Lt. Robert H. Wilson poses astride his downed P-38. (J. Bradley McManus)

my plane and we dressed his arm and made him lie down for a while so he would be able to get to Hanna's B-17 which had just landed. We took some snapshots of the wreckage while he rested. Looking at the hole from which he had crawled we wondered how he possibly managed to emerge from it. The cockpit was under the snow. He had evidently smelled the puff of smoke which I had seen, thought the ship was on fire and started fighting his way out. In the process he had cut his arm on a piece of jagged glass which had pushed into the cockpit. He had cut himself free of the parachute and opened the side window thus giving himself about eight inches clearance to squeeze out of the plane. He had come out feet first. To this day I cannot understand how he escaped with such a minor injury.

"The next morning by the clock we sawed the props off one of the engines on Staples' B-17 in order to start the engine to keep the batteries charged as we radioed for help. After some experimentation we finally devised a stove out of what had been an oxygen cylinder with an exhaust manifold as a flue. Burning oil in this makeshift stove finally allowed us to warm the B-17s to the point where we had a very cozy home in which to live. During the third day at our hotels(the B-17s were dubbed Hotel Big Stoop and Hotel DoDo) we were spotted by Navy Lieutenant Atteberry in a Catalina Flying Boat. He dropped us bed rolls blankets, canned foods, whiskey, magazines and, during our stay, flew over us at frequent intervals to keep our morale up. The whiskey which was dropped was used for medicinal purposes though we could have used some for other purposes had it been there. We used it to wash the cut on Brad's arm. We also took a swallow each day and had only one cold among us during the eight days. Several transports flew over us and two B-17s and two B-24s dropped us supplies. We had radio contact with Bluie West 8 and were able to receive radio (music and news broadcasts) from Iceland and from London, thus it became a question of waiting for our rescue party to reach us. (Lieutenant Harry Smith was somewhat more descriptive in referring to the dropped rations:

Staples' B-17 crew outside their "Shack". (J. Bradley McManus)

Robert B. Wilson poses on the wing of the B-17 as the stranded flyers prepare to trek back to civilization. (J. Bradley McManus)

"I ain't in favor of the flavor, make damn good dog food though.'"

"We had prepared sleds to be pulled by us and had our most prized possessions on the sleds. Then we saw the party coming over a rise in the snow. They were Naval Lieutenant Fred Crockett, Sergeant Don Shaw and Corporal Harry Kent. All of them were expert skiers; Crockett had been with Admiral Byrd on his Antarctic expedition in 1929. After 8 days on the icecap we were ready to get back to some semblance of civilization. We took food for six days and what we thought we could carry with us on the long hike we knew was ahead. At about 10:30 PM we started from the crashed ships on the longest walk I have ever taken. We were eager and full of pep and did not realize what lay ahead.

"Early on, some sleds gave way to backpacks which weighed as much as sixty pounds, even after some lightening by discarding whatever could be discarded. We usually walked in thirty minute stretches broken up by a five minute rest. As the time dragged on we were becoming more and more weary. Our walking time decreased to fifteen minutes with three minute rests. We were constantly perspiring from the exercise of walking and there was a bitterly cold wind blowing at all times.

"We arrived at the shoreline of a small bay where there was a small open boat that had transported our rescue party down from Bluie East 2, a distance of about sixty miles. The open boat, which we were more than happy to see, took us to the Coast Guard Cutter **Northland** which had picked its way among the bergs to reach us. We boarded the cutter, cleaned ourselves up and were treated to a wonderful dinner and enjoyed very much the day and a half cruise to BE-2. The crew and officers aboard the cutter treated us royally.

"We were picked up by the PBY's and flown back to BW-1. During the trip we flew directly over our ships on the icecap. They were arranged in a semi-circle with Brad's ship in the center and slightly ahead. If we had landed in formation we could not have arranged them in any neater pattern. We remained at BW-1 for about two weeks during which time we slept about 18 hours out of the 24 and ate the rest of the time. From BW-1 we were flown back to Goose Bay on August 14th. Later that same day we were flown back to the States to Presque Isle. We all rushed to the mess hall of course. It was a very good feeling to be back in the States living like Americans."

None of the P-38 pilots who survived the icecap landing ever returned to the 94th or the 1st Fighter Group. Each was reassigned after his return.[2]

Consider for a moment flying the North Atlantic in an early model P-38 in the summer of 1942. Those planes were cold particularly at altitude and looking down at an iceberg studded North Atlantic made them even colder. Navigation in the subarctic is not easy and navigation aids were practically non-existent with the exception of a few relatively short range homing beams at, for example, Bluie West 1. In the subarctic radio communications were often garbled or blotted out entirely at random intervals. Little was known of winds aloft or even of weather systems given the dearth of reporting stations in the far north.

In view of these difficulties the pilots who made the crossing were fortunate. In a way, so also were those like Bob Wilson who were forced to land on the icecap. They at least were over something solid. Had they been forced to come down in the water it would have been an entirely different story. None would have survived. There is no question that the powers that be were prepared for some losses. In fact we lost no pilots and only six P-38s and two or more B-17's (at least one of the B-17's landed long at Goose Bay). It could have been much worse.

As this history is being written the six Greenland P-38's are still in the news. A recovery operation has been ongoing for several years with partial success. The planes have been located over 250 feet down in the snow field. At least one P-38 has been recovered with Brad McManus there to record the event.

Over the remaining days in July the 71st squadron and the balance of the 94th left Iceland for England (Scotland), flying on the wing of a navigator B-17. The 27th squadron air echelon remained in Iceland. So

The Greenland Expedition Society camp site, July 1992. (J. Bradley McManus)

ordered by the War Department to protect against German long range aircraft. On July 7th the 27th relocated from Reykjavik to Keflavik a new field and small town not too far away. Elmer Hartman, a 27th pilot who kept a (probably) illegal diary describes the 27th stay in Iceland as unrelieved boredom for the most part. His flight was one of the first of the squadron flights to arrive in Iceland and throughout most of July he anticipates leaving for England only to be told on July 26th that the 27th would be in Iceland for the duration of the BOLERO Operation. In his words, "I was ready to stick my head in a prop." In truth the stay in Iceland was pretty grim to judge by Hartman's diary. Bad weather, iffy food, not much flying and very little to do added up to very long days. Hartman's diary notes that he flew for an hour on August 17th; "First flying time in 22 days!"

Looking down the recovery shaft. (J. Bradley McManus)

July 15, 1992, fifty years to the day after the massed forced landing on Greenland's ice cap, the Greenland Expedition Society basks in the glow of their recovery achievement. (J. Bradley McManus)

Recovered nose section of a P-38. Brad McManus is on right. (J. Bradley McManus)

P-38s of the 1st Fighter Group at Reykjavik, July 1942. (Unk.)

On August 14th the drudgery ended with a bang. Lieutenant Elza Shahan of the 27th and Lieutenant Joseph Shafer of the 33rd Squadron (flying a P-40) shot down a Focke Wolfe 200 Condor, one of the German long range reconnaissance aircraft. This was the first enemy aircraft shot down by American forces in the European Theater and provided an instant shot in the arm and morale booster. Relieved by the 50th Squadron, 14th Fighter Group the 27th finally departed Iceland August 22nd, landed at Stornaway for "weather reports and tea and cakes," then on to Prestwick, Scotland and finally to Wellington, England and a reunion with the ground echelon. The 1st Fighter Group, in its entirety was, at long last, in England.

Overall the BOLERO operation, between June 1942 and January 1943, delivered 366 heavy bombers, 150 mediums and 183 P-38s to England. The winter of 42-43 presented major problems. An accident rate of 2.9% in September rose to 5.8% in October and continued to rise. All operations ceased in December 1942 and prohibitions against P-38 travel came much earlier. For that reason only the 1st and 14th Groups took part in BOLERO. The 82nd Group, both planes and men, was transported to England by ship. In retrospect only the war emergency could have justified the risks involved in the over water mission. With very few exceptions the pilots had no experience in over water flight. They depended upon transient, half completed facilities and upon very uncertain weather forecasts in an area where weather constituted, under the best of conditions, a major hazard to flying.

Bluie West 8, the field in far northern Greenland, was originally to be part of a great circle route from California via Great Falls Montana, across Canada representing the shortest distance between the California aircraft industry and England. A German submarine intercepted and destroyed much of the needed equipment and winter weather put an end to the plan.

1. BW-1 was located on the Southwest Coast of Greenland and BW-8 was on the same coast but further north.

2. Robert H. Wilson and Carl Rudder later served with the 14th Fighter Group in the MTO. Harry Smith joined the 82nd Fighter Group also in the MTO. Brad McManus and Robert B. Wilson saw combat in Europe with the 364th Fighter Group, and Dallas Webb was also assigned to the ETO with the 55th Fighter Group. All were P-38 units. All six pilots survived the war.

3

ENGLAND - 1942

After landfall in Scotland, some food and good Scotch whiskey, the 94th and 71st Squadrons began the process of organizing in England. Colonel Stone assumed command of the group replacing Major Zahn. Headquarters, the 71st Squadron and the 27th ground echelon located at Goxhill near Hull. The 94th went on to the RAF base at Kirton-in-Lindsey in Lincolnshire. Finally the air echelons, two of them at least, were reunited with their entire ground contingents. And none too soon. Some of the ground echelons were beginning to feel pressed into the British army what with English food, "God Save The Queen" and assembling for parade under British orders. The complete squadrons soon put a stop to that and conditions reverted to United States fighter squadrons albeit on English bases. For the 94th at least, Kirton-in-Lindsey with its barracks and concrete runway would be the last permanent base the squadron would see for a long, long time.

Also on the field at Kirton was the Polish RAF 303 Squadron and training began immediately. The Poles were at Kirton for a rest period. They were veteran fighters; several had ten or more victories and their experience was invaluable. Training films, gun camera film and aerial maneuvers were shared and were most useful. By coincidence the Polish squadron was the same one visited by Glenn Hubbard on a fact finding mission before our entry into the war. Jack Ilfrey described the Poles as merry fellows with an immense capacity for whiskey. They were also extremely rugged individuals and deadly fighters.

On August 19, 1942, although the group was not yet operational, the 94th suffered its first loss in England. Flying as a part of a high altitude practice mission, around 20,000 feet, Lieutenant Charles Oakley went into an uncontrollable dive from which he could not recover. Why he went into the dive remains a mystery; his wingman, Lieutenant Pennington, could not stay with him and pulled out although barely. Oakley was killed when his parachute did not open in time.

Uncontrollable dive characteristics, a result of compressiblity, were to plague the P-38 until dive brakes were finally installed on later models but not until 1944. The phenomenon of compressibility was a continuing problem for the P-38 and other high performance fighters. In brief, in high speed dives, usually from high altitude, the air speed over the wing could reach the speed of sound when the aircraft speed was considerably less. The resulting shock wave rendered the plane uncontrollable. The speed at which this phenomenon occurred was known as the critical Mach number.

As the squadrons settled into the English countryside the men began to make some contacts with the English people and understand that we are indeed two countries separated by a common language. In time we began to ride bicycles, comprehend but feel uneasy with a batman[1], drinking tea and, in extremis, eating orange marmalade and doing without coffee. Dewey Prince, of the 71st squadron, decided that he rather liked the English hot tea, "if they wouldn't put milk in it. The only tea we drank at home was iced tea and we certainly wouldn't put milk in that."

The 27th squadron, relieved after its additional month in Iceland, located at Wellington and began in earnest with a conditioning program, intelligence reports, aircraft identification and combat air maneuvers with units from a Spitfire squadron.

Dewey Prince recalled one unfortunate example of guardsmanship while the squadrons were still at Goxhill. One of the men assigned guard duty on a rainy night began his tour wearing a raincoat. The rain stopped so he took off his raincoat and hung it on the pitot tube of the nearest P-38. An hour or so later, returning on his

Flagpole at Kirton in Lindsey, July 1942; RAF flag, top, American, Polish for Kosciusko 303 squadron. (Tom Cleland, Lockheed tech rep via Ilfrey)

L - R Wing Cmdr. Drobinsky, C.O. of the Polish Wing, John Zumback, C.O., 303 Sqd., a 12 victory ace, and Glen Hubbard C.O., 94th Sqd. (Jack Ilfrey)

tour he saw what appeared to be someone standing under the wing. Too dark to make out who or what it was. There was no answer to his repeated challenge so he fired his Thompson submachine gun on full automatic. He shot one raincoat and the back end of one P-38 full of holes.

Late in August 1st Group Headquarters and 71st Squadron moved to Ibsley. The 94th also moved south to a field outside of Ibsley, just north of Bournemouth on the Southern Coast of England and the group was declared operational. Early in September the 27th moved to Colerne near Bath and also became operational. The move to Ibsley was not without its problems. On arrival, in the rain, the 71st diary records as many as four moves in one night, from one barracks to another, before getting a few hours of uninterrupted sleep. Morale was not particularly high at this point.

On September 1st the 94th and 71st made the first all-American equipped mission, a fighter sweep into occupied France. Not an enemy plane was seen. Sweeps such as this together with some B-17 escort missions were the beginning of what was to become the massive Eighth Air Force raids of late 1943 and 1944. During September and October the group flew a total of twelve missions over occupied territory, as Jack Ilfrey phrased it, "all the while expecting fierce opposition

94th pilots at Kirton in Lindsey. Standing L - R Ilfrey, Pennington, Kille, Neale, Widen, Rimke, Hubbard, Harman, Sutcliffe, Williams, Wilson. Kneeling, Hille, Murdock, Starbuck, Pringle, Jennings, Roberts, McWherter, Malzahn. Nearly half would be casualties by mid-1943. (Jack Ilfrey)

Briefing the troops on arrival at Ibsley, England. (Lenhart)

and never encountering it." By early September rumor had it that the group would be departing England. Elmer Hartman describes one rumor to the effect that they (the 1st) would not be in England long but would embark on a distant journey. Speculation was on two possible destinations; Siberia, where the Japanese were expected to attack and the Egyptian front.

Despite several flights into enemy territory the next casualty occurred in England. As a part of the training exercises Lieutenant William Pennington, a 94th pilot, was directed to make a low level strafing pass at one of our anti-aircraft gun positions near Ibsley. Due to some miscalculation he pulled up too late and mushed into the ground. Obviously enemy gunfire would not be the only cause of casualties in this war.

About this time Captain Clyde Gabler, group S-2, became a near casualty and experienced a strange military attitude. Gabler and an enlisted man crawled into a crashed, inverted fighter, freed and extracted the Rhodesian pilot amidst pools of leaking petrol. Needless to say there was a great risk of the whole thing exploding, and the two were recommended for a citation. The recommendation was returned with the notation that, "since the plane did not ignite no award was appropriate." Of course, had the plane ignited any award would have been posthumous.

The time spent in the Bournemouth area was clearly practice time as well as operational time but practice for what? While the pilots were making the first tentative passes into enemy territory the ground crews were constructing wooden shipping crates, moving equipment by hand and gassing the planes using five gallon jerry cans; this last a long, arm and back wearying procedure. During these last few weeks in England there was an apparent reluctance to commit the P-38s to anything more than very brief overpasses on the fringe of Fortress Europa. The Lightnings and German fighters eyed each other from some distance but encounters were rare. The P-38s seemed to be deliberately held back from either bomber escort or fighter attacks on military installations within easy reach of the long range fighter. Of course, in the summer of 1942 the heavily armed B-17 was considered self sufficient by many in the Air Corps. Schweinfurt was well in the future.

The 27th was a little behind schedule due to its stay in Iceland. However, during late September any number of planned raids (B-17 escort) were called off because of the weather. Finally on October 9th all of the squadrons joined with Spitfires in escorting B-17s to Lille, France. Elmer Hartman's diary notes, "saw approximately 12 enemy ships but they kept at a safe distance." On one of these sweeps we did lose a man however. Lieutenant William Young, a 71st pilot, went down over France; just how was not certain.

During September and October there was even more guessing about where we might be going. Some said France although that seemed unlikely in view of what had happened at Dieppe just a few months ago. A few mentioned North Africa although no one could figure out how the planes would get there. After the BOLERO

Operation the answer should have been obvious. Elmer Hartman had it pegged pretty well noting in his diary on October 12th, "We will go to Gibraltar and then to Egypt via Africa". Pretty close to the truth.

In mid-October an order was issued for each man to be armed. By that time officers and the first three grades had been issued sidearms-the .45 caliber automatic. The rest were issued a rifle, as it turned out either a Springfield or a Garand. These men, for the most part, were aircraft specialists and knew very little about the use and care of a Garand.

Late in October the guessing came to an end and people began to move out. Clyde Gabler describes Colonel Stone's briefing " We are shipping one-third of each squad on a separate ship so that if we lose one, we will still have two-thirds of each squad." Captain Gabler was to command one-third of the ground echelon. The first ground contingent left Ibsley by train on October 25th and arrived in Glascow the following noon. The Glasgow group boarded an English troopship that afternoon. Other contingents leaving Ibsley went by train to Liverpool and boarded ship there. The three ships involved, the *Mooltan*, the *S.S. Orbita* and the *Empress of Canada* departed from ports on the west coast of England and Scotland. The ship transit was reasonably pleasant - if a troop ship voyage can be called pleasant - with the apparent exception of the *Orbita.* On that ship according to reports, the distinction between officers and enlisted ranks was carried to extremes. The enlisted men were crowded, cramped, had little food, most of it bad, i.e., breakfast of well watered tea, a hard bun and kippers that had seen too much sun. To add insult to injury, the ship's crew breakfast of bacon, eggs and coffee was carried through the enlisted mess while they were enjoying their weak tea repast. Definitely there were no happy campers on the Orbita and with good cause. The ship, incidentally was scheduled to be scrapped in 1935 but was somehow still afloat.

On November 6th all air echelons moved to Chivenor near Lands End. Departure was imminent - or would be when the rains stopped.

1. A British enlisted man who served as a sort of "Man Friday" to officers.

4

OPERATION TORCH

Clyde Gabler describes the ocean transit from England to North Africa: "We were several days at sea before the opening of sealed orders. We were on our way to North Africa. Soon the word was passed to all the troops with several men collecting on bets they had made as to our destination."

"The 26th of October found the Mooltan out on the Atlantic Ocean travelling round in big circles for eleven days. The weather was mostly good with only occasionally rough seas. Our unit was assigned a space on deck for an hour each day. There we did exercises in order to keep in shape for the landing we knew lay ahead....

"On November 6, 1942 the convoy stopped circling and headed toward the Straits of Gibraltar at increased speed. We passed through at night and could see the lights of the city of Tangier twinkling in the distance. The next day we were escorted by some destroyers. The convoy was attacked by German U-boats and the destroyers lobbed depth charges which shook our old ship to the keel."

"Our destination was east of Oran at a place called Arzew. Our ship anchored there about a mile off shore on 8 November 1942 together with other ships of the convoy."

After waiting on deck in full combat gear most of the day all units were told to go back to the cabins and holds. Debarkation would be the following morning. Finally at first light on November 9th the 2nd echelon of the 1st Fighter Group boarded the landing craft and went ashore on the beaches at Arzew.

Ray Ainsley, with the 94th ground echelon, describes his experience:

> "On November 10th we stopped (anchored) just off the coast of St. Leu, just east of Oran. The first troops got off early the next morning and we got off about 4:00 PM. We went down rope ladders into small landing boats. Blood was all over the boat I was in. They said a First Sergeant had been hit in the head while getting off. About 40 feet from shore the boat ran aground and they sent us into the water. I was about fourth to jump in and wade to shore. I had a field pack in one hand and a bed roll in the other with a life preserver around my neck. About the time I jumped, a big wave came along and hit me in the rear. I went down on one knee but kept my pack and bedroll dry, however I did get my watch wet. It never did run afterwards. By the time we got ashore and halfway organized it was dark. We started walking up the hill in a staggered, dispersed formation. Every once in a while you could hear a rifle crack. We had three days of C rations and they were good after the terrible English rations on the ship. We had to exist on Limey rations for nineteen days. We marched through St. Leu about 10:00 PM right through downtown then down the road about two more miles and found ourselves lost. It was dark as pitch, raining and what a mess. Finally they decided the best thing to do was try to sleep till daylight. I had a bad night sleeping on a pile of rocks soaking wet. The next morning we went over to an Air Corps camp about a half mile away. There we encountered our first Arabs. We gave them candy and cigarettes and they stole more than we could guard. One little guy stole one of my pipes. Water was very scarce and we had to purify what little we could find. The pills made it taste terrible.
>
> "November 11th We got our first action. The Germans had captured two British Spitfires and came over strafing us. I ran for open country but before I could get very far that Spitfire was coming directly at me. I turned lengthwise to his approach and got by OK. I will always remember the Arabs running across the field in a bunch with their robes waving like a bunch of ghosts."[1]

Dewey Prince of the 71st says, "We went ashore at Arzew. I was picked to lead a squad of 12 men. I positioned my men in a single line behind me and prepared to move down a street which ran along the waterfront. Another squad was positioned across the street with a third squad positioned behind mine. Sergeant Loren Eck was in front of me. We had only moved a short distance when the shooting started. I yelled for the squad to take cover and tried to see where the shooting was coming from. I looked back to see if the men in my squad were all right. One was sitting in the street holding a Garand rifle in his lap and a clip of bullets in his hand. He looked up and said 'How in the hell do you load this damn thing?'

"When the shooting stopped we assembled our men and started down the road again. The shooting started up again and we took cover again. When I looked for cover the only place I could find was between two boats and Sergeant Eck had gotten there first. Eck was a big man I am a small man so part of me was sticking out in front of the boats. About that time Eck said 'Damn it they got me.' I told him 'It's just your imagination.' He said 'Imagination hell, look.' He pulled back his collar and had a wound in the flesh just above his collar bone. The bullet must have passed over me and hit him because I was standing directly in front of him. Eck got bandaged up, put his pack on the other shoulder and we walked about seven miles to St. Leu to join the rest of the squadron. I don't think he ever even checked in at a hospital with his wound."

Another group of 71st people had an interesting and somewhat lonesome landing. Lieutenants Peterson, Bowyer and twenty eight enlisted men were waiting their turn for the landing barge when two other groups intervened. By the time the 71st group landed, or waded ashore, it was dark and the rest of the 1st Fighter Group was nowhere to be seen. Lieutenant Peterson could find no one so his group bedded down for the night in a beachfront cottage. They ate their emergency rations (better than the *Orbita* fare) and went to sleep to the sound of heavy guns firing. The next morning they discovered that their beachfront abode was surrounded by explosive shells. The group was next directed toward the city of Arzew ostensibly to meet with the rest of the 1st Fighter people. On arrival they were first scattered by snipers and secondly discovered that they had been sent in the wrong direction. Extricating themselves from the snipers, by sniping back, they retraced their steps to the cottage they had started from. By that time it was getting dark again so they decided to settle in there again. The next morning they set off in the opposite and hopefully the correct direction. After another couple of miles they finally found the rest of the group. In short order everyone headed again for Tafaraoui.

The trek to the airfield at Tafaraoui was a curious mixture in Clyde Gabler's words. "Some of our super mechanics had found an old truck at a vineyard, started it up, went to the head of the column, filled it with men, took them to the airfield and came back for more." This leapfrog operation plus a couple of more trucks finally got everyone to the airfield. Two other observations parallel Captain Gabler's words describing the Tafaraoui trek. "They managed to lay hands on some ancient French buses and, as John Holt puts it, 'it was the knowhow of the 94th mechanics that got the antiques running and transported the whole group, or most of it, to Tafaraoui.'"

Don Anson says they commandeered a truck, a trailer and a bus, loaded them all so that the whole procession looked like something out of the Grapes of Wrath, but it got the job done. The urgency, of course, was the expected arrival of the air echelon. The 27th was first to arrive landing at Tafaraoui on November 13th accompanied by sporadic artillery fire from the nearby hills. This was quickly put to rights by the 1st Armored Division.

The story of how the three air echelons reached Algeria takes us back to Lands End, England. The squadrons, with the 27th leading, departed England beginning on November 9th. The plan sounded simple. Each two flights totalling eight P-38s would team up with one B-26 for navigation, fly across the Bay of Biscay, down but well off the Spanish coast, cut diagonally across Portugal to landfall on the coast of Spanish Morocco, turn left through the Straits of Gibraltar thence

Tafaraoui, Algeria, November 1942. Twelfth Air Force Headquarters displays both U.S. and French flags. (Lenhart via Ilfrey)

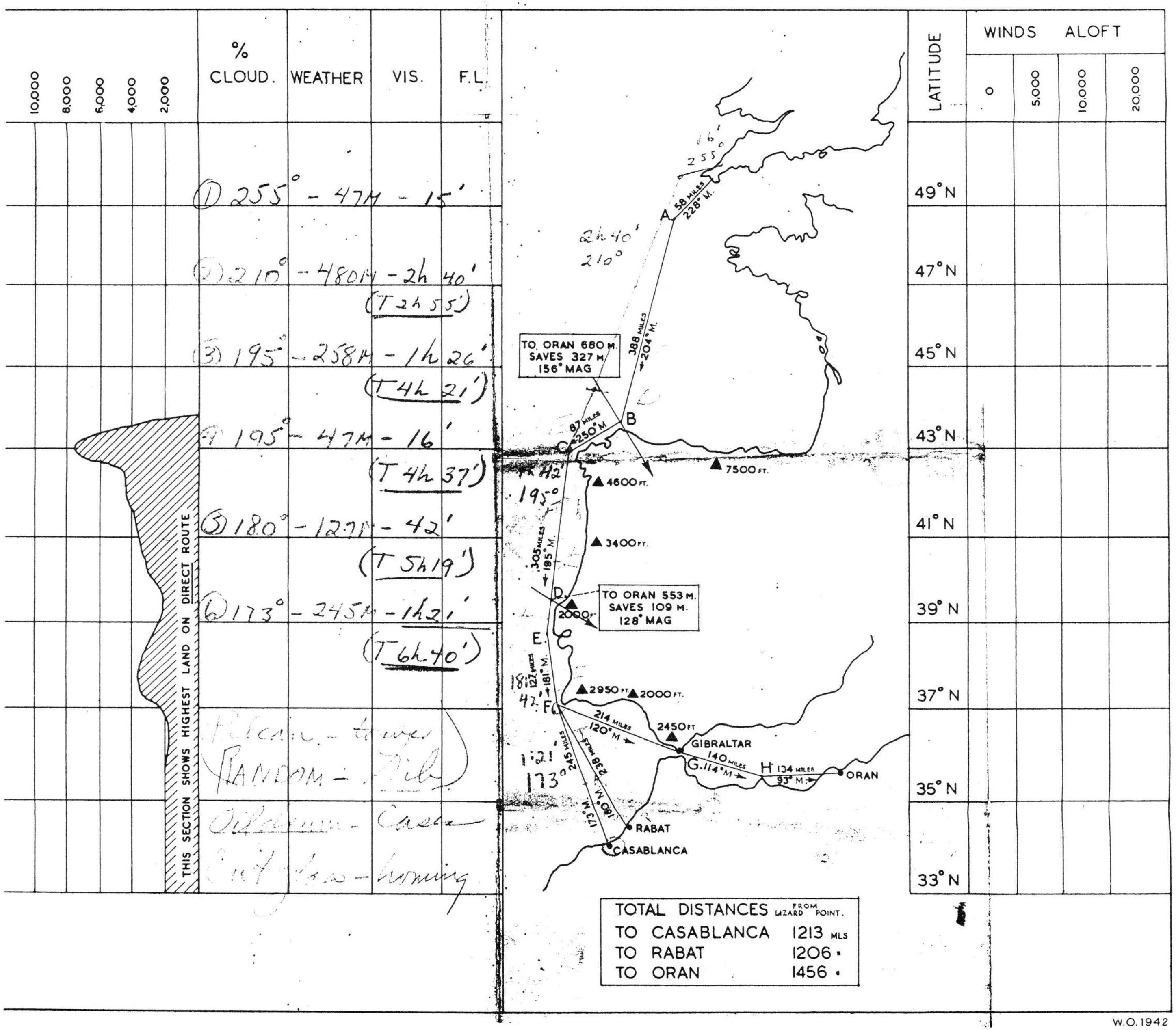

Route map of Lieutenant Sam Sweet, 27th Squadaron, England to North Africa. (Sally Sweet Stangl)

to Oran. Gibraltar was English, of course, but was to be used only in dire emergency, i.e., no gas. The plan essentially copied parts of the BOLERO mission substituting B-26s for the B-17s, somewhat easier flying because the B-26 was faster.

The trip to Oran turned out to be a mixture of the routine, or what passed for routine, to the unusual to the extraordinary. Elmer Hartman of the 27th recorded his transit in his diary:

> "Finally got off the ground on Sunday the 15th of November. Left Chivenor at 0800 hours and the trip took a total of eight hours, a long time considering being confined to a cockpit with no movement. Weather was very good considering such a long trip, a distance of 1500 miles. Flew at 500 feet from England to Spain to avoid being picked up by German detectors along the French coast. When we reached Spain we gradually climbed to an altitude of 5000 feet and flew along the coast until we reached Portugal. From the northwest point of Portugal to the southeast we travelled a straight course. We made certain not to travel over Spain which is technically neutral but definitely pro-Nazi. Portugal was very much as I had imagined it, lots of orchards and vineyards and rolling hills. Houses are mostly red-tiled roof and single stories. Weather forced us down to a low altitude halfway across the country and we crossed several towns. I could see the people standing in the streets waving at us. Somewhere along this route Lieutenants Wolford and Chenoweth left the formation to go on their own. Wolford made it alright but Chenoweth hasn't as yet showed up. It is thought he went down in Spain and is now a prisoner there. [Chenoweth apparently crashed in the ocean and was killed.]
>
> "From Portugal we flew straight to the northern tip of Spanish Morocco, adjacent to Gibraltar. Unfortunately we were not informed of the anti-aircraft guns installed there and which the Spanish are not in the least reluctant to use and we therefore suffered several good scares. Flak greeted us in all directions. One burst occurred just off my right wing tip but caused no damage. Here we left the B-26 and proceeded on to Oran. Landed at a field close to town which is filled with bomb craters. Had a hell of a time landing and taking off. Field was covered with dust, several planes crashed into bomb craters and one Spitfire crashed into a plane on the ground."

Bob Neale of the 94th crash landed some miles west of Oran, several had to refuel at Gibraltar, but the majority managed to land in the midst of chaos at Tafaraoui some fifteen miles south of Oran. Of those whose flight was interrupted Jack Ilfrey's experience was probably the most unusual not to say bizarre.

Here is Jack's account taken from his book, **Happy Jack's Go Buggy:**

"There was a low overcast that morning. Good weather had been predicted all the way to Oran but I felt an uneasiness I was unable to analyze. It's good we didn't know the higher ups had prophesied only 50% of us would reach Oran and as the minutes passed the flight was proving to be not quite as difficult as we thought it would be. It had been made plain at briefing that we were not to land at Gibraltar unless it was a real emergency, as the Rock was jammed tight with personnel.

"Suddenly I felt a jolt and by the time I realized what had happened my right engine went out. One of my long range belly tanks containing 150 gallons of gasoline had fallen off. Automatically I switched to another tank and the right engine caught again. But one fact was crystal clear. I had lost 150 gallons of gasoline.

"I didn't want to go back to England. I wanted to stay with my outfit. I got out my maps and charts and after a few quick mental calculations decided I had enough gasoline to get me as far as Gibraltar and dismissed all thought of going back. I was not going to miss out on the operations in Africa, and I was not going to be separated from the gang. And, being honest, I have to say I wasn't going to miss the big adventure ahead - Africa. Maybe I was thinking of lions and tigers. I don't know.

"We kept on flying and when the B-26 which was leading us dodged a thunderhead I had the feeling we were getting off course. I flew for another hour and began to get worried. After another hour I began to get alarmed. My gas was getting low as we had used a great deal of fuel in dodging thunderheads. We were pretty far out to sea. I knew that much and I knew we were flying in a southwesterly direction. I decided to leave the group as my gas wouldn't last much longer (All of this, of course, was under radio silence so there was no discussing the situation with the squadron leader or anyone else).

"I turned southeastward and hoped to hell I would run into the Spanish or Portuguese coast. The clouds were dispersing and when the sun came out I hoped I would see the coast. And in a few minutes I saw it when I was flying at an altitude of 8000-10000 feet. Again I got out my maps and saw that I was close to a point on the coast which was marked the Spanish-Portuguese border. I then flew parallel to the coast and after a short time realized I would never make it to Gibraltar. My gasoline was almost gone.

"There are no words to describe the feeling you have when you are eight to 10,000 feet in the air and know your gasoline is going to be exhausted at any minute. The best description I can give is that you have a sort of paralysis with part of your mind still functioning between what to do next and hoping for something you know is not going to come true. One minute I decided to bail out. The next minute I thought I'd crash land on the coast. While I was still wavering I came to the mouth of the Tagus River,which, according to my maps ran 20 or 30 miles into Lisbon We had been poorly briefed on what to do if we were forced down in either Spain or Portugal. I vaguely remembered someone saying the Portuguese were more friendly than the Spaniards and might be bribed to get you out of the country. I knew too that it was the duty of every pilot forced down in enemy or neutral territory to destroy his plane and equipment. Still not knowing definitely what I would do, I turned inland and headed toward Lisbon.

"And presently I saw a beautiful airdrome lying just outside the city of Lisbon - long beautiful runways and big administration buildings. Everything looked inviting and seemed to beckon a welcome, and, without another thought, I put my wheels down, circled the airdrome and landed. When I had completed my landing roll, I saw six men, mounted on horseback, galloping out to meet me. The horsemen looked like something out of a picture book. Big, plumed hats, sabers, pistols and gay trousers, and for a split second I thought of the changing of the guard at Buckingham Palace. Gesturing wildly, the men motioned for me to taxi my plane toward a building which turned out to be the administration building, and, meanwhile, I was hurriedly tearing up maps, papers and throwing them to the wind, and with more frantic gestures, the guards told me to stop on the apron just in front of the building.

"When I killed my engines I looked up and saw people coming from every direction toward my plane. Some of the people in the crowd gave me the "V" for victory sign, but most of them just stared, and by this time the horsemen had surrounded my plane.

"I stepped out of the cockpit and looked down at the crowd and I felt like a stranger to myself. It was almost impossible to believe I was in Portugal, about to become an internee. When I inquired if anyone spoke English, a young fellow who looked sort of ex officio came up and said he did. As I remember the conversation it went something like this:

"You are an American, yes?"

"I certainly am an American." (Why I said' certainly' I don't know, unless it was subconscious pride coming to the front.)

"The Portuguese smiled and said: 'Yes I thought so. I see the star on your plane. It is the first American warship I have seen.' (A distinction of sorts here, the first P-38 to be forced down in Portugal.) Then he added, still smiling: 'In trouble?'

"My reply was serious enough: 'Yes, I am in trouble. I need some gasoline. Is it possible for me to get any?'

"The fellow gave a short laugh. 'Come with me and I will take you inside.' As I walked away I glanced back and everyone was staring at me and at my plane in amazement. I felt a sudden alarm about my airplane and I asked the Portuguese official if anyone would bother the plane and he spoke to one of the guards who commanded the people to stay away from the P-38."

After a period of unremitting confusion and intense interrogation, Ilfrey was ushered back to his aircraft by a Portugese Air Force pilot who requested a cockpit check, all this while the Portugese were refueling the plane. The intent left Jack feeling helpless and forlorn, when providence reared its head.

"Suddenly I heard a familiar, faint sound, and then a louder noise, and I looked up and saw a lone P-38. It was in trouble and making preparations to land. The six guards on horseback dashed off and people started running toward the end of the ramp to watch the landing. The P-38 was putting it's wheels down and was landing on one engine.

"I held my breath and all of a sudden it struck me: 'Ilfrey, what the hell are you waiting on?' I threw on all the switches and while I was doing this, the Portuguese pilot realized what was about to take place and he tried to reach inside the cockpit and turn off some of the switches. But lightning had hit and I had already turned over my left propeller and the engine started; and the pilot, who was sitting on the left wing, lost his hat when the propeller created a terrific wind, which sent what was left of the crowd scurrying away. When the pilot started trying to get a better grip on the plane, I started my right engine and threw on lots of power, creating a bigger wind than ever and managed to keep the airplane still. My idea was to blow the pilot off the wing, which I did in another second or two, and when I looked back - after the Portuguese had cleared the plane - I saw the officials holding onto their hats while many others were chasing their headgear, and, believe me there were plenty of Portuguese hats flying in the air. I threw the canopy shut, gave the plane full power, and without looking back went straight across the field, disregarding runways and everything else in my path."

Jack Ilfrey landed at Gibraltar with most of his belongings, but his helmet was still in Portugal. He had flown into a storm of international complications. Portugal was upset, Washington was angry and the Colonel commanding at Gibraltar was absolutely furious. For a while it seemed likely that Jack and his plane would be returned to Portugal and interned. Fortunately reason eventually prevailed and he was allowed to proceed to the doubtful luxury of North Africa.

The pilot in the other P-38 landing at Lisbon as Jack was taking off was less fortunate. This was Lieutenant James Harman of the 94th. One engine was feathered and, as he was landing, he saw another (Ilfrey's) P-38 taking off. He expected a friendly reception and a quick turn-around. Instead the irate Portuguese practically clapped him into irons. After some 4 months he was finally smuggled out of Portugal and rejoined the 94th in North Africa. To say Harman was irritated at Ilfrey would be putting it mildly. It seems strange from this distance but when Harman rejoined the 94th someone at the General HQ level raised the same kind of fuss that had been raised when Jack Ilfrey flew out of Portugal. There apparently were some who thought Harman should be returned. Hubbard and the 1st Group HQ thought that was ridiculous and the whole matter died.

The arrival of the entire 1st Fighter Group at Tafaraoui could have been far more difficult than it actually was. The first squadron to depart from England, the 27th, left on November 12th. The cease fire agreement with the Vichy French had been signed only the day before on November 11th. (The Armistice Day of the war to end all wars).

Prior to and during the first few days of the North African invasion the intentions of the French were unknown and nearly unpredictable. During the early stages of Operation Torch, French resistance in some areas was fierce and stubborn. The labyrinth of Vichy French and Free French politics left one group of French regarding assorted other units of French Army, Navy or Air personnel as traitors.

Yet a cooperative French Morocco/Algeria was crucial to any hope of Allied success against the Afrika Korps. The French cease fire - it was not termed a surrender - meant the Allies, including the 1st Fighter Group, could move toward Tunisia hampered only by logistical and weather problems, which were bad enough. Had we first had to fight the French, the Arabs, Berbers and God knows who else while Germany was pouring troop and air reinforcements into Tunisia, the outcome in North Africa could have been very different indeed

The stay at Tafaraoui (where the mud was deep and gooey) was fortunately brief. Facilities, what there were, were makeshift and overcrowded. People were sleeping on everything including the ground and floors. The field was badly overcrowded with aircraft parked everywhere. The 71st orderly room was located in a discarded French bus-probably the first portable orderly room in AAF history. The German long-range bombers could have wreaked havoc and it's surprising that they didn't.

1. The official record says an American unit was strafed by two 31st Fighter Group Spitfires which were misdirected and shot down but doesn't say specifically who did the misdirecting. Communications between Allied air and ground units were extremely poor at this stage of the war.

Capt. Jim Harman's P-38F-1 in Lisbon, Portugal in Cross of Christ Squadron markings. (Jack Ilfrey)

5

FIRST BLOOD

In late November 1942 (beginning on the 17th) the 1st pulled out of Tafaraoui and moved some 40 miles east to Nouvion. Located in a level valley and still in the Oran area, what had been a grassy field at Nouvion was even muddier than Tafaraoui had been. Pup tents and mud were the order of the day. In November the North African winter was beginning; a dismal prospect. Cold rain and a clinging mud made even the routine chores hard to deal with. The mechanics and line people adopted woolen caps and all the heavy clothing available. Facilities at Nouvion became progressively worse if that was possible. Clinging mud, living in shelter halves, powdered eggs, refueling planes from 55 gallon drums, things could not have gotten much worse. However, once again after a brief stay, the group, or at least some of its parts were on the move. The 94th squadron was ordered to move to Youks les Bains to reinforce the 14th Fighter Group. Departure began on November 25th. The squadron barely got off the ground and then had to leave the gear down, because both the gear and wheels were so mud caked. A short time later, in early December, the 71st Squadron - at least the air echelon - was ordered to Maison Blanche, an airfield just outside of the city of Algiers.

At Thanksgiving the 71st had a 'Bring your own' dinner which worked quite well. It was a break from stew-stew-stew in the rain-rain-rain. Incidentally, a few days after the departures, the 27th Squadron remaining at Nouvion was unable to provide B-17 escort for a raid on Bizerte. The mud was so bad the P-38s couldn't take off. (**Fighters Over Tunisia** p. 72) However in early December part of the 27th was sent to Algiers to reinforce the 71st. The balance of the squadron moved to Maison Blanche on December 13th en route to Biskra. Shortly after arriving at Nouvion the 27th got themselves crosswise with the authorities. Assigned an afternoon escort mission all three flights buzzed the field after takeoff. They returned to a serious dressing down by Major Garman.

Raids on Bizerte and Tunis were difficult, particularly from Nouvion. The distance, nearly 600 miles meant six hour or longer missions with the usual sweating out the gas situation. The German response to TORCH led to a major transfer of fighters and bombers into the theater. As early as November 4, 1942 Luftflotte 4 (stationed on the Russian Front) gave up a fighter group to the Mediterranean. (**Luftwaffe**, Murray). The Germans and Italians had concentrated their air defenses at all weather fields at Sidi Ahmed at Bizerte and El Aouina at Tunis supported by bases at Sfax, Sousse and Gabes. All of these were supported and reinforced from enemy bases in Sardinia and Sicily.

While still operating from Nouvion the 27th squadron claimed the initial 1st Fighter Group victory of WW II.

On November 30, 1942 the squadron had been in North Africa about two weeks and had had no contact with the enemy. Two previous forays had been turned back by bad weather. Captain Joel Owens led a flight from the 27th Squadron on a November 30th mission to Bizerte, Tunisia. Other members of the flight were Marcus Linn, a former RAF pilot, Henry Smith and Harold Mendenall. Owens describes the engagement:

> "Today sixteen P-38s of the 27th were to escort 18 B-17s of the 97th Bomb Group to Bizerte. The target was the North Quay at the Naval Base. The

94th Sq. personnel about to move up to forward bases via C-47 transports. (Lenhart)

timing was ....take-off when it gets light enough and get home before it gets dark. Since we were situated about 500 miles from the targets we generally arrived over the target at a predictable time. Today the time over target was 1215. An open air briefing was conducted by Major Weltman. As the pilots gathered around he spread out his maps on the hood of his Jeep.

"The bombers were to proceed to an initial point (IP) midway between Tunis and Bizerte at 24,000 feet. At the IP they were to turn left on the run to the target, followed by another left turn after bomb release. The four flights of P-38s were to rendezvous with the bombers over our base and follow at the bombers pace at around 170 MPH, above and to the "up sun" position in relation to the bomber group. The B-17s were in their usual box formation of six planes in stepped down Vs of three planes each. The three boxes were deployed in a loose V. The up sun box was about 200 feet below the lead box and the down sun box about 200 feet above the lead box. As we approached the IP, the P-38s would increase their speed and climb to their assigned altitudes. The squadron leader and eight aircraft were the close escort whose job it was to S back and forth at five to seven hundred feet over and near the bombers. One flight of four was the support force whose mission was to be ready to protect up sun and prevent the close escort from being 'bounced'. The top cover was provided by A flight. We were to be about 1200 to 1500 feet above the support flight and were allowed quite a bit of leeway insofar as our position in relation to the bomber force.

"The squadron was off on time and the mission proceeded as briefed. I was leading White flight with Junior (Lieutenant Linn) on my wing. Mendy and Smitty were the second element. The weather was clear when we took off but gradually worsened as we went along. The cloud deck was solid as we passed Algiers but as we approached the IP there was a hole in the clouds over the target. The B-17s bombed through the hole. There was some flak but no fighter opposition. The bomb group was off to my right and about 2000 feet below as they turned left and headed for home. The other P-38 flights were close to the bombers as the entire formation passed below us.

"At that moment I saw a single-engine fighter as he was climbing up through the hole. I started after him by making a steep spiral dive. I had the presence of mind to check to make sure the guns were hot and all the switches were on. We started our dive about 26,000 feet and had lost about 9,000 feet. As the speed built up the plane became nose heavy and the controls started getting squirrely as we approached the edge of compressibility. As we neared the enemy plane I identified it as an Me. 109E. I was closing much too fast to make a good attack and was pulling a lot of "G's" as I started to fire from 150 yards at an angle of around sixty degrees above and behind him. I fired a short burst and, wonder of wonders, managed to hit him in the left wing and the aft part of the fuselage. The German pilot did a half role to the left and popped down into the clouds. I pulled up into a steep climbing turn to the left and headed westerly to rejoin the formation. I looked around to see how A

Lt. Lewis Murdock, 94th pilot "at home", Youks les Bains. (Jack Ilfrey)

The French garrison at Youks Les Bains poses before a 1st Fighter Group P-38. The unit, dressed mostly in gear that dates back to World War I, included a few colonial troops. (Ken Sumney)

flight was doing and Mendy was the only one in sight. I asked where the rest were? Junior said something but all I could make out was, 'Over!'. Smitty called that he and Junior were together but Linn had an oil pressure gauge that was bothering him. They were just above the cloud deck heading home.

"At that time I saw another Me. 109 about a quarter of a mile to our left and slightly below us. He was climbing on a southwesterly heading that would put him in position for an attack on the bombers. I put 'the balls to the wall' and we started after him. I managed to get directly behind and slightly below him as we closed to a range of 100 yards, my pulse rate matching my air speed, as we passed through 25,000 feet at 220 indicated air speed. I was getting extremely tense, edgy, nervous, apprehensive or just plain scared as I recalled the words of an RAF pilot who said, 'The most dangerous time, chaps, is when you've got your pipper on the bloke.' I had closed to about 200 feet when the 109 banked slightly to the right. I thought, 'Oh,oh this is it. You had better do what you are going to and right now!' I fired a short burst, and had him bore sighted. Then I clamped down on the gun button until there was an explosion in the fuselage and the tail blew off.

"The Messerschmitt started tumbling and a great deal of smoke was coming from the midsection. Mendy hollered, 'You got him Rum, you got him.' I replied, 'Yeah, I got him.' Then I uttered the immortal words that are universally used by fighter pilots to this day 'Let's Get The Hell Out Of Here!'"

On return to base after a six hour forty-five minute mission, part of it nursing a wingman on one engine, Owens put on a real buzz job including a slow roll on the deck. Landing he could scarcely wait to tell the world about his feat. In the midst of his story Colonel Garman and Major Weltman pulled up in a staff car looking like a pair of thunderclouds. Weltman's first words were approximately You $#!**@(*&^%&&!! What The Hell Are You Trying To Do? Colonel Garman used more refined language but his displeasure was clear, "Joe you know the policy, What's the idea?" Once the aerobatics were explained they were both all smiles and wanted the details. After a session with the interrogators Captain Owens was credited with one Me. 109 destroyed and one damaged.[1] Crediting victories was easier said than done using the rules at that time, i.e., one witness, no one could have seen the enemy plane hit the ground in a partly cloudy sky from twenty odd thousand feet. The gun camera in the early P-38 was mounted so close to the guns that the film was often a blur when the guns fired.

1. *Fighters Over Tunisia* by Shores, Ring and Hess says that "Otto Reuss of IV/JG 53 was wounded in combat with P-38s while on a routine test flight over Bizerte.....and Lieutenant Horst Wunderlich of II/JG 51 failed to return from a mission in the same area."

CHRONOLOGY OF AIRCRAFT USEAGE BY DATE OF APPEARANCE

DATE	TYPE
May 1942	P-38E, F
Mid -1943	P-38G
Jan. 1944	P-38H
Mar. 1944	P-38H-5, J-15
Aug. 1944	P-38J-25
Nov. 1944	P-38L

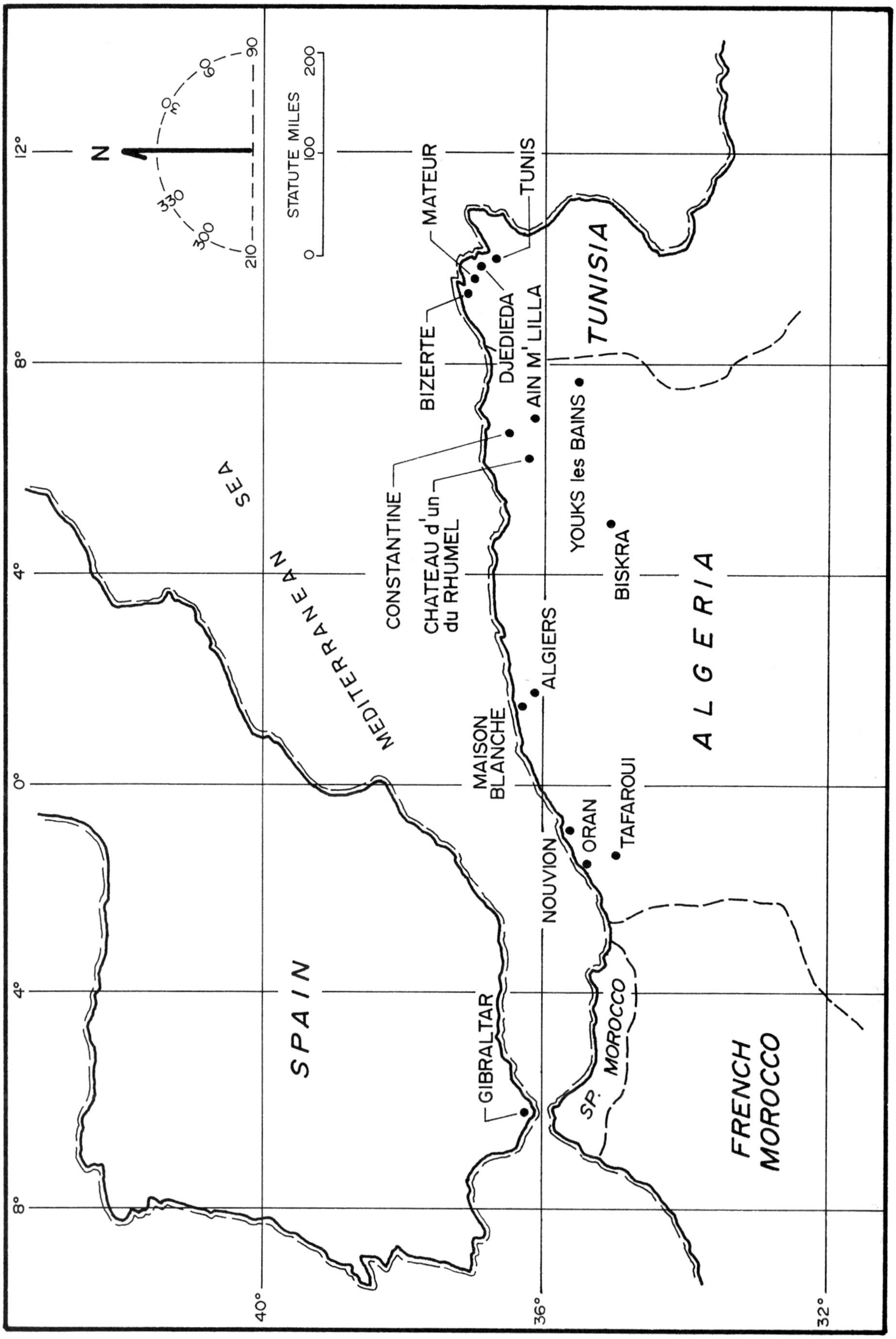
MEDITERRANEAN
SEA
SPAIN
ALGERIA
TUNISIA
FRENCH MOROCCO
SP. MOROCCO
GIBRALTAR
NOUVION
ORAN
TAFAROUI
MAISON BLANCHE
ALGIERS
BISKRA
YOUKS les BAINS
CONSTANTINE
CHATEAU d'un du RHUMEL
AIN M'LILLA
DJEDIEDA
BIZERTE
MATEUR
TUNIS
N
210
300
330
0
30
60
90
STATUTE MILES
0
100
200
12°
8°
4°
0°
4°
8°
40°
36°
32°

6

THE BATTLE FOR NORTH AFRICA

On the eve of departing Nouvion Major Glenn Hubbard, commander of the 94th, reported a squadron of 33 officers and 276 enlisted men. The 94th left Nouvion, destination Youks les Bains, on orders to reinforce the P-38 equipped 14th Fighter Group. The 14th, with one squadron still in Iceland, had been engaged principally in front line troop support and found heavy going due to piecemeal commitments for troop support. Reinforcement was overdue. [1]

With this move the 1st Fighter Group began hopscotching eastward across North Africa. At times the three squadrons would be together and then they would operate separately until finally rejoining at Chateau d'un du Rhumel.

Youks les Bains, the site of ancient Roman baths, is located somewhat north of Tebessa in eastern Algeria. The airfield was a grassy, soon to be muddy field beside a railroad track. Youks was quite close to Kasserine Pass and within striking distance of Faid Pass, Gabes and other German strongholds in Tunisia.

The 94th arrived at Youks on November 28th; at least the planes and pilots did. The ground contingent was en route via a variety of conveyances. The 94th would operate from Youks until December 6th. From then until the 14th planes, pilots and some support personnel would operate from Maison Blanche just outside Algiers. On December 15th the 94th would reform in Biskra where all of the 1st Fighter Group would reassemble. Major Hubbard's record of events records the transit from Nouvion to Youks les Bains.

> "November 28, 1942. The combat echelon consisting of twenty-three officers and their aircraft departed from Nouvion and arrived at Youks les Bains. All pilots arrived safely. The air echelon remained at Blida awaiting the arrival of French transports and spent the night in the hangar. All officers and enlisted men present. The train echelon, consisting of 2 officers and 10 enlisted men departed from Nouvion by rail with 10 cars of equipment. They departed during heavy rain and mud.
>
> "November 29,1942 The combat echelon and aircraft now at Youks les Bains were without ground crews and the pilots had to service the aircraft. On this day the squadron went on its first operational mission. Six P-38s took off at 0635 hours to strafe the Gabes airdrome. The flight was led by two pilots of the 14th Group. The formation made two passes at the airdrome. A water tower was hit. The P-38s then strafed a train about fifteen miles north of Gabes. Two passes were made at the train and the engine was hit. At this point one P-38 left the formation and headed for home due to lack of fuel. On the return trip the flight encountered two Me.110s. Captain Roberts (Newell O.) and First Lieutenant Ilfrey (Jack) shot down one of the planes and it was seen to crash. The other was shot down by 14th Group pilots. The flight landed at 0910 hours. All pilots returned safely.
>
> "At about 1245 hours a flight a four P-38s took off to escort nine B-26s. The objective was to bomb Gabes. The bombing was observed to be accurate. No enemy aircraft were observed. The flight landed at 1430 hours. All pilots returned safely.
>
> "Another flight of eight P-38s took off at 1300 hours. The objectives were Mateur, Bizerte and Tunis. Two Me.109s were observed at 10,000 feet but were too high to intercept. A Ju.88 and probably an Me.109 were intercepted and shot down. The Ju.88 was claimed by Captain Robert E. Williams and the Me.109 was claimed by Lieutenant Umphrey and is claimed to be a probable. Lieutenant Forsythe (Paul J.) became separated from the formation and shot down one Ju.88 and probably an Me.109 before he was forced to crash land. The airplane was completely demolished but Lieutenant Forsythe escaped injury. The remainder of the flight landed at 1500 hours. All pilots returned safely.
>
> "The air echelon, now at Blida, remained there awaiting the arrival of French transports and spent the night in the hangar. All officers and enlisted men were present. The rear echelon arrived at Maison Carre about 1700 hours and spent the night on the train. All officers and men were present. The echelon remained here overnight due to Movement Control uncertainties as to the destination of this echelon-and, one might add, a chronic lack of transport."

The lack of motor transport was so bad throughout the Twelfth Air Force that during the Casablanca conference a special sea convoy was laid on which subsequently brought 5000 trucks from the States. Both of these echelons were a very long way from Youks les Bains at this point.

While the 94th was having an exciting time of it at Youks and wondering if the entire squadron would ever get together, the 27th and 71st were flying, or attempting to fly escort missions. There was an escort mission to Bizerte on November 29th, six and one-half hours flying from Nouvion only to find 100% overcast in the target area - mission abandoned. However, on December 3rd the 27th got into it with a vengeance. Escorting B-17s again to Bizerte the 27th was jumped by a flock of Me.109s. Elmer Hartman split-Sd from 20,000 feet and escaped but reported Lieutenants Long, Everett, Drysdale and Hester missing with at least three 109s downed. Hester was in fact wounded in the melee but made it back to base. The others did not.

The next day, again over Bizerte, the 71st had its turn when Captain John Eiland downed an FW 190, however the 27th lost two more pilots in the same area when Lieutenants Lawrence Pace and Hubert Black went down.

Major Hubbard's entries for the 94th Squadron continue:

> "November 30, 1942. The combat echelon at Youks les Bains, still without ground crews, carried on servicing the aircraft themselves. Twelve P-38s took off at 0930 hours to escort nine B-26s. The objective was Djedeida. The bombing was observed to be accurate. No enemy aircraft observed. The flight was down at 1140 hours. Four P-38s took off to escort nine B-26s. The objective was Gabes.The bombing was observed to be accurate. One B-26 was hit by flak and was escorted by two P-38s. The B-26 crash landed in enemy territory and the location was fixed. The formation landed at 1500 hours. All pilots returned safely. Lieutenant Umphrey took off later with an A-20 to pick up the crew of the B-26. The mission was successful. Twelve P-38s took off at 1425 hours. The objective was Djejeida. Two Me.109s were observed by them but were not engaged. The flight landed at 1645 hours. All pilots returned safely.
>
> "The air echelon still at Blida were awaiting the arrival of the French transports. The night was spent in the hangar. All officers and enlisted men were present. The rear echelon was still at Maison Carre awaiting orders from Movement Control. The night was spent on the train. All officers and enlisted men were present."

By this time all three echelons of the 94th were getting more than fed up with their situations, what with sleeping in hangars, sleeping on trains and, of course, the pilots flying and servicing their own aircraft. On the 1st of December the various roadblocks came loose at once. The air echelon arrived from Blida in an assortment of transports and later in the day the rear echelon arrived as well. It goes without saying that the pilots were overjoyed to see both contingents. Raleigh Johnston recalls a gourmet meal experienced during one of these train rides, including cream of wheat, vienna sausage and tomato juice.

Gassing the planes at Youks was a back breaking job. The gas in five gallon cans was dropped off in front of each plane. The cans had to be heaved up on the wing, stabbed open with a screw driver or equivalent and poured into a tank, then tossed off the back to be picked up again by the truck on its return. The motto for the pilots was don't go around unless it's vital. That circle could mean emptying three or four more cans of gas.

Despite worsening weather, activities at Youks were stepped up. Our raids were hurting the Germans but their supply lines were considerably shorter than ours and we were not keeping up. On December 2nd a four plane flight was off at 0645 to strafe Faid Pass. The flight continued on to the Sfax-Gabes area where six Me.109s were sighted on the ground. Two were destroyed by Captain Newell Roberts. Lieutenant Jack Ilfrey shot down an airborne Me.109 and Lieutenant Richard McWherter downed two more. Lieutenant William Lovell downed yet another and shared one with Jack Ilfrey.

In a mission post mortem Captain Newell Roberts, the flight leader, noted that after strafing Faid Pass the flight continued on to the city of Sfax on the sea coast. Newell described flying down the main street of Sfax at zero altitude and seeing German aircraft in buildings along both sides of the street. The street was being used as a runway with buildings converted into hangars. Continuing into the Gabes area the Roberts flight tangled with a flight of Me.109s In this melee Lieutenant Jack Ilfrey's plane was hit and his right engine knocked out. He proceeded back to base at the best speed he could manage on single engine-about 275 mph- escorted by Lovell and Roberts. Someone shouted "Weave, Jack, weave." All made it back but not before Ilfrey's P-38 was well peppered by gunfire from at least two enemy aircraft. After landing Jack's plane had a total of 268 holes including several large ones from anti-aircraft fire. A cannon shell had demolished the radio and another had hit the propeller of the left (good) engine. Jack got the P-38 back to Youks but it was never to fly again. Hugh T. "High Tension" Jones recalled an interesting remark from Jack telling his crew chief to thank the radio man that he could hear someone yelling, "Weave, Jack," before the radio was hit. H.T. said after hearing that how could I ask, "Why am I here?"

Activity at Youks continued at a high pitch. On that same day twelve P-38s from the 94th provided escort for mediums on a raid to El Aouina, a large German airfield at Tunis. Still later the airfield at Djedieda was strafed yielding more ground claims. Lieutenant Umphrey claimed an Me.109, Glenn Hubbard and Victor Giles shared a Ju.87 and Clarence Rimke and Lewis Murdock shared a second Ju.87. On the 3rd of December Gabes was strafed once again with results that were less fortunate. Lieutenant Jack Longseth was hit, crashed and killed. Captain Williams was hit by flak, crash-landed and became a POW.

At Youks the 94th was actually in a different part of the Twelfth Air Force designated the XIIth Air Support Command. This was a front line infantry support

Jack Ilfrey's P-38F-1 made it back from Sfax on its good right engine. The left was knocked out by Me. 109s. Note the outbound hole just ahead of the open cowling. The plane was also hit by flak and sustained some 260 holes. Ilfrey on top, left, amd Lt. Bill Lovell, top, right, inspect the damage. (Ken Sumney)

command comprising fighters (P-38s, P-40s and some P-39s) and light bombers (A-20s). This contrasted with the XIIth Bomber Command which included escort fighters (P-38s), heavy bombers (B-17s) and mediums (B-26s and B-25s). There was no separate fighter command.

While, by early December the 94th was reasonably intact once again, the food situation had not improved one iota. The food wasn't American and it wasn't quite British. It wasn't plentiful and it certainly wasn't good. Basically two meals, breakfast and supper included powdered eggs, a GI can of beans and hardtack washed down with tea. Water was scarce except for the Roman baths at Youks which did at least afford personal cleanliness on occasion.

Some days after the ground crews arrived at Youks Jack Ilfrey returned from a mission with his plane well shot up. The cables in one boom were severed by anti-aircraft fire. These cables control elevator and rudder movement. Jack knew the boom had been hit but wasn't aware of the extent of cable damage. As Herb Green recalled it we didn't have facilities to remove

Another view of Ilfrey's "Texas Terror". Second from the left is crew chief, Roy Silvers. The entire left tail boom was removed to inspect for control cable damage. (Jack Ilfrey)

the boom and replace the cables. What cables we had were scavenged from wrecked P-38s. John Holt had the idea to overlay the cables wrap them with safety wire and solder the whole thing. This was not looked on with great enthusiasm but after thinking it over it was decided to make a test run. Someone rounded up two pieces of cable, the ends were overlapped about three inches, wrapped with safety wire and the whole business soldered. To test it a bomb hoist was brought up and one end of the cable tied to the frame and the other to the hoist winch. With all hands watching this crazy scheme the crank was started. After a few turns-crack-the cable parted but the splice had held. Salvage another plane. The full cable was spliced and while the end result looked good it was a little short so they stretched it a little by tugging it between two jeep bumpers. Ignorance is bliss, Ilfrey kept flying knowing none of this and later shot down two FW 190s operating from Biskra in his reconstructed P-38. He actually found out about the cable stretching at the Langley group reunion in 1984.

The move of the 94th to Maison Blanche was not intended to recombine the three squadrons but was probably caused by the weather as much as anything else. Fourteen P-38s departed a wet and muddy Youks on December 6,1942 for a very brief stay at Maison Blanche. At this point the weather began to give everyone a break. The group had been pressed hard in their first days of combat action against an experienced and well equipped Luftwaffe. By the same token the Germans were beginning to feel the effects of our bombing and strafing of their all-weather fields at El Aouina, Gabes and Sfax.

In mid December the 94th suffered one of those inexplicable losses. On an escort mission Major Hubbard, Captain Roberts and Lieutenant Umphrey spotted an Italian Cant Z-1007 bomber at low altitude over the Gulf of Tunis. All three shot it down, however Lieutenant Umphrey's P-38 continued down in a gentle glide, made what amounted to a belly landing in the Gulf then slowly sank into the sea. He made no move to escape.[2]

Officers' mess, Biskra. Murdock (L), McWherter (R). (Jack Ilfrey)

The one year anniversary of December 7th passed almost unnoticed, certainly unheralded. And yet just a year before, a part of the 94th had been landing at El Paso on its way to the West Coast with the remainder of the group just returning to Selfridge. The events during the succeeding year seem almost incredible: West Coast patrols, the BOLERO Mission, England, Operation TORCH and finally Algiers and the Casbah. It was hard to believe yet this was only the beginning. In the short span of two to three weeks the 1st Fighter Group had been introduced to hand to mouth, cold, muddy operations and a cram course in combat flying. In late 1942 the Americans, although experienced pilots, were combat novices. Until Youks les Bains and the Bizerte missions, most had not heard a shot fired in anger. Suddenly they were up against an experienced Luftwaffe flying first rate aircraft. The Me.109 and particularly the FW 190 were the equal of the P-38 and in some respects superior. In addition to the Germans, enemy forces included the Italians who were somewhat unpredictable.

The Americans, with practically no combat experienced pilots, had to rely on their brief training with the Polish RAF squadron and their various associations with RAF pilots. All of this made for a fast learning curve. The learning process is best described by the developments in formation flying during escort missions. After trying line abreast and echelon formations the so-called "finger four" evolved and was used till the end of the war. At first twelve planes, three flights of four, would make up a squadron mission. These used an alternate weave pattern to maintain a decent airspeed and sky visibility yet not outrun the bombers, particularly the B-17s. Late in 1943 the sixteen plane box formation was developed still using the cross weave pattern. The four plane finger four formation was simply a doubling of the basic two ship element of leader and wingman. However, too often this became a flight leader, wingman and element leader with a 'tail end Charlie' trying not to get left behind in a kind of crack-the-whip effect.

The P-38 units were, in fact, developing the technique of bomber escort as well as their own fighter tactics. Before the war and even in the early days of WW II many believed that the heavy bombers could protect themselves. The British method of night bombing did not use fighter escort so the American fighters had little precedent for escort techniques.

Joel Owens reflected on the role of fighters flying in support of the heavy bombers as the role was being developed early in the war:

> "The fighters ceded the initiative to the Luftwaffe as we committed to the role of close escort. We never seemed to get too close to suit the bombers. At times we put a flight in line abreast directly beneath the lead box of bombers. The idea was to compensate for the lack of foreward firepower on the B-17Es and to deter the enemy fighters from making sustained head-on attacks. We

experimented with new formations and tactics almost daily in an effort to increase our effectiveness and eventually settled on a loose finger four formation for the individual flights, with the squadron right down on top of the bomber formation. Within the flight, if operating alone, the two elements would fly abreast and spaced widely enough to allow either element to turn inside and protect the other. The key word was protect. Even when our squadron fighter strength dwindled to the point that on December 15th, 1942 two of us wound up escorting eighteen B-17s on a mission to Tunis, we STAYED WITH THE BOMBERS!

"In the beginning we had little or no experience in bomber escort. The RAF had been of little help. They had scads of experience in intercepting enemy forces but the fighter units had done very little escort work. In fact the RAF squadrons differed greatly in their concept of the proper squadron formation. The choices were many and varied: line abreast, line astern, box, weavers, finger four - you name it, they had it. The Luftwaffe had the most experience in daylight escort but their record in the Battle of Britain was not all that great. Anyway, they were unavailable for consultation.

"Our formations and tactics clearly were not designed to attack and destroy enemy fighters. They were designed to prevent the bombers from being shot down by sustained attacks of enemy fighters. To that end I believe we were successful since the majority of the bomber losses in the Tunisian campaign were due to anti-aircraft fire or mid-air collisions. We didn't know it at the time but we were facing some of the Luftwaffe's elite pilots. JG 53 had at least a dozen pilots with twenty or more victories. The top gun in JG 51 was Anton Hafner who would go on to score 204 victories before he was shot down. The tactics we used and the discipline which enabled us to carry out our mission in the face of such formidable opposition are a credit to leaders such as Glenn Hubbard, Ray Rudell, Willie Weltman and Ralph Garman. They started from square one!"

The role of the fighters in bomber escort would continue to evolve as the fighter squadron formation changed from the very close escort to a three flight weave and finally to a four flight weave, flying more and more complex protective formations high over larger and longer bomber streams.

Early in December the 71st squadron got its first victory in the form of an Me. 109 shot down by Captain John Eiland during a mission escorting B-26s to Bizerte. In Eiland's words the action took place as follows:

"Lieutenants Sgroi and Schoenfeldt, in the rear of the formation, reported FW 190s pealing off to attack. First Major Rudell and next Lieutenant Muldoon took their elements back to give the rear section some help in driving off the 190s; but as Lieutenant Muldoon turned he was jumped at once

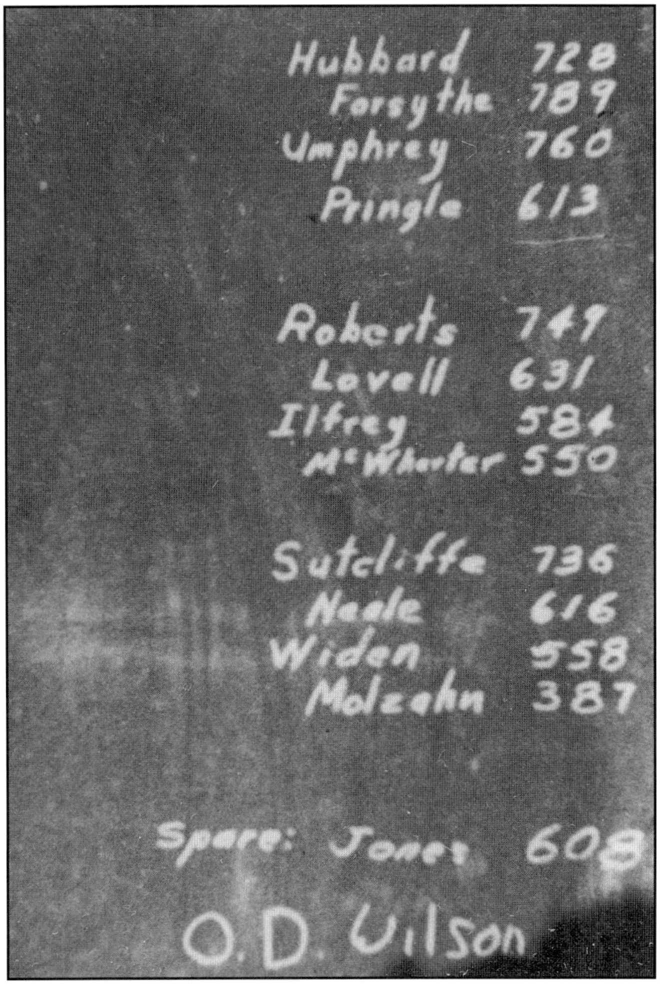

Mission assignment, 94th Biskra. Six of these men were later killed in action, one taken prisoner and one wounded. (Jack Ilfrey)

by two Me.109s who had apparently been waiting up in the sun. He called over the radio asking, 'How about some help back here? 109's on our tail!' And so I took my section back to give some help.

"In the meantime Lieutenant Sgroi, who had first reported the enemy aircraft attacking, had outrun his 190 and Major Rudell had driven off the plane on Lieutenant Schoenfeldt's tail. But since Muldoon had been jumped, the trouble was far from over. As I turned back I saw that Lieutenant Muldoon had apparently outmaneuvered his opponent but there was a 109, bigger than hell, flying formation with Lieutenant Russell and shooting the shit out of him. It was beautiful to watch. They were racing all over the sky; half rolls, steep banks, climbing turns, abrupt dives - everything in the book but spins and still the two planes stuck close together. But now I saw a third, a 109 trailing behind the first one. Two planes on his tail!

"As I closed into the action, Russell was just pulling up in a steep left turn and I think by that time, the first German plane, the one flying Russ' wing, saw me coming so he peeled off and headed for home. The second one however, decided he would close in and make the kill; so he made a pass and ended up so that I was able to get in position for a little of the same thing he was dishing out. At that time he was 100 feet off the water and pulling up gradually into a slow shallow right turn. I brought my sight on him and cut loose with all five guns. He was an easy shot, just a slight deflection to the right, almost dead ahead. I followed him slowly around, pouring it on with long bursts. At length, all my cannon was expended and about half my machine gun ammunition. At the finish of the last burst I saw pieces of his fuselage near the canopy fly off and figured,'That's all brother!'

"He did a half roll and came back under me so that I lost him for a moment and was unable to get in any more shots. And then I saw him again, heading down toward the water, on his back, with the prop slowed up considerably, glycol streaming down the fuselage and a light burning smoke coming out of the engine."

"At length he righted the plane to a normal position and it gradually skidded into the water. It did not look as thought he were making a landing because he made no attempt to level off; just splashed and skidded. The plane streaked across the water for a ways, slowly losing speed and sinking as it went along. I saw no evidence of the pilot escaping, presumably he was hit and sank with the plane."

"I pulled back up to join the formation, doing no further investigating since he looked pretty well done in, and besides, there were still four enemy planes above."

The stay at Maison Blanche was short. After about a week all three squadrons pulled up what few stakes there were and moved once again, this time southeast to the desert for a somewhat longer stay. Prior to this however, the group experienced some promotions and reassignments. Colonel Garman became commander replacing Colonel Stone who was assigned to Twelfth Fighter Command. On December 5, 1942 Major Weltman moved from the 27th to become group Exec, with Captain Joe Glenn succeeding him as 27th CO. Captain Walter Roney became 27th Operations Officer one of the few non-pilots to hold this position.

Roy Silvers bartering for eggs, Biskra. (Jack Ilfrey)

Biskra was a prewar oasis resort in the Sahara that encompassed some 3000 acres and included what had been resort type hotels. Located southwest of Tebessa and well southeast of Algiers, its airfield was a large area of hard packed sand. Life among the date palms was quite a change from the rains, mud and mountains further north in Algeria.

At Biskra the 1st Fighter Group once again assembled and functioned more as an intact fighter unit, however the group by no means enjoyed Biskra by itself nor did it function as it would later in the war. More often than not the squadrons operated independently which, of course, meant smaller numbers against the Luftwaffe. Also operating out of the oasis were the 97th and 301st Bomb Groups flying B-17s and the 17th Bomb Group flying B-26s. This made for a very trafficy area, and a tempting target for German bomber crews. The mechanic's work at Biskra was somewhat easier in that gas was now available in fifty-five gallon gas drums equipped with a pump. No more tossing cans of gas up onto the wing. There were still no work stations however, so engine repair work had to be carried out standing on a gas drum while someone else patrolled the perimeter carrying a machine gun or carbine.

It is worth remarking that prior to Biskra the three squadrons of the 1st Group had really been accustomed to operating individually. This was true during the stateside maneuvers. In California the squadrons were in different locations, the 27th was on its own in Iceland and the group was in divided locations in England and in North Africa prior to Biskra.

At Biskra the 94th gave up troop support as its primary role and joined its sister squadrons in bomber escort work and longer range strafing attacks. The escort missions included escorting B-17s or B-26s to Tunis, Bizerte or ranging out over the Mediterranean to targets in Sardinia and Sicily. And the losses mounted. On a B-17 escort mission to Bizerte on December 18th George Sutcliffe was hit and downed. He suffered some burns but made his way back via Algiers. Norman Widen was shot down and became a POW and Clark Jennings was shot down but returned aided by British Commandos and continued flying. Victor Giles was hit and killed. All of these were 94th pilots. With the 27th Squadron Henry Smith was declared a prisoner of war and Captain Theodore Runyon was missing in action on December 14, 1942. Lieutenant Lawrence Butler of the 71st was killed at Maison Blanche on December 1st.

There is an interesting sequel to the Norman Widen loss. He was shot down by a German pilot named Anton Hafner mentioned earlier by Joel Owens. Hafner went on to become one of Germany's most decorated fliers. He was killed in 1944 after amassing over 200 aerial victories. Prior to his death he had requested that his brother locate the American he had shot down on

December 18, 1942 and give him one of his (Hafner's) decorations. This required a long search to identify and locate Widen. Hafner knew that Widen had survived being shot down as the two had met briefly in North Africa before Widen was transported to Germany and imprisoned. Hafner's brother accomplished his task in 1960. (From **Fighters Over Tunisia** p. 106 and **Kings Cliffe Remembered** Vol. 7 No. 4).

Christmas Day at Biskra was a day like all the rest. An uneventful escort mission was flown but no turkey dinners were consumed. Ray Ainsley's diary describes Christmas dinner as, "Limey steak and kidney pie, mostly kidney". An unconfirmed report says that later on that day someone brought in a keg of muscatel which was consumed well but not too wisely. The 71st squadron was in the midst of moving to Biskra and spent a divided Christmas. Some were in Biskra, others on a French train and the remainder still at Maison Blanche. All had one thing in common: no turkey. On the day after Christmas on an escort mission to Tunis, Jack Ilfrey shot down two FW 190s, however, Lieutenants' Milton Sahl and Richard Jones were lost, both killed. Lieutenant Robert Sauer of the 71st bagged a Ju. 88. Thus ended 1942, certainly a year of many trials and many errors but overall a distinct improvement over a year ago.

The new year of 1943 was barely started when one of the group's most improbable escapes occurred. Lieutenant J.C. Harrison (Bugs) Lentz, a 94th pilot flying escort near Tripoli, was hit and downed. He was captured unscathed and imprisoned in Tripoli. After some time he and several other prisoners, English and American, about a dozen in all, were put aboard an Italian submarine to be transported to Europe and ultimately to a Stalag Luft in Germany. A day or so out of Tripoli the sub surfaced at night close to a British destroyer which immediately opened fire. Amidst the turmoil on the sub the prisoners, or most of them, forced their way topside. The sub soon sank and most of the crew and all but two of the prisoners were lost. The destroyer crew nearly went into shock when the first two survivors pulled from the sea turned out to be British and American. Lentz was eventually dropped off at Malta, outfitted with some British Army togs and flown back to the 94th. A part of the story which may be apocryphal has it that months later when Lentz was back in the States he received a bill for those British uniform items.

January continued with a mix of escort and strafing missions. Escorting B-26s to Tunis on the 1st, the 71st claimed two Ju. 52s shot down by Lieutenants Sauer and Meldrum Sears. On the next day Captain Joe Glenn destroyed an Me.109 but was wounded in the process. Glenn was the commanding officer of the 27th Squadron and was replaced by Captain Joel Owens. In the same encounter Lieutenants Mendenall and Weil each claimed an Me.109, however Lieutenant Henry Smith was shot down and taken prisoner.[3] On January 11th on an escort mission to Bizerte Lieutenants

94th Operations, Biskra, J.C.H. 'Bugs" Lentz leaving. (Jack Ilfrey)

William Lovell and William Wilson of the 94th were killed. Lieutenant Joseph Smith of the 71st was killed on a fighter sweep in the Gulf of Gabes on January 12th, reportedly machine gunned in his parachute. On the 14th of the month in a combined attack against the airfields at Sousse and Sfax the 71st lost Lieutenants Salts and Meyer and the 27th's T.H. Runyon bailed out wounded west of Gabes and was captured.

On January 15th Lieutenant McWherter was killed in a taxiing-takeoff accident. The billowing dust at Biskra so obscured visibility that the two moving P-38s could not see each other and collided resulting in the death of McWherter, who was taking off. The other pilot (Kelly?) apparently survived with only minor injuries. Lieutenant Harold Lentz, a 94th replacement pilot and not to be confused with 'Bugs' Lentz, was in line to takeoff and thinks the dust was started by takeoffs of the B-17s located across the field from the Lightnings. Harold achieved some immediate notoriety in both the 94th and the group by identifying an island off the coast from Bizerte as a battleship. He said it led to quite a bit of close questioning and the posting of a cautioning notice on the squadron operations hut warning new pilots against such island identifications.

On January 18th escorting B-17s to Castel Benito the group mustered thirty-three P-38s. Lieutenant Burton Weil went down to become a prisoner of war, however Lieutenants Stevens and Sullivan each claimed a Macchi 202. Thirty-three aircraft for the group works out to little more than 10 aircraft for each squadron-less

Ernst-Wilhelm Reinert, an ace of II/JG 77 scored 51 victories in Africa. Lt. Burton Well, 27th Sq, was one of his victims. Reinert survived the war with the rank of Lt. Col and 76 air-to-air victories. (James V. Crow)

than half the normal level. This was typical of the aircraft supply for each of the squadrons of the three P-38 groups although the 14th was by far the worst off of the three.

On first arrival at Biskra foxholes were the order of the day. Sleeping under the date palms was such a change from the mud at Youks and Nouvion that it even seemed romantic - maybe for a day or two. In time some of the resort hotels were taken over and assigned to the pilots, five to a room. Just take your bedroll and find a spot on the floor, as there were no beds. It was in the hotel Transatlantique that Ernie Pyle interviewed several pilots. In one of his columns he mentioned spending some time with five pilots. Within a week there were only three. The five pilots were Ilfrey, McWherter, Neale, Lentz and Shelton all of the 94th. Lieutenant Harold Lentz provides a further description of the hotel rooms: "They measured 8 x 12 with 15 foot ceilings and a 4 x 8 door; one window above eye level, lighted by a single bulb hanging from a drop cord and activated by a rotary switch at the door, no heat. Furnishings included an army canvas cot and an orange crate."

Ilfrey interview with Margaret Bourke White, Life Magazine photo-journalist, Youks les Bains. (Jack Ilfrey)

Margaret Bourke White, a photojournalist for Life Magazine also spent some time at Biskra early in January interviewing some of our people. Later in the month things were enlivened by the arrival of a USO troupe including Kay Francis, Martha Ray, Carol Landis and Mitzi Mayfair. The girls put on a terrific series of shows and pleased all hands.

The ground crews at Biskra continued their ingenious ways. From spare bits and parts they fashioned the only radar unit at Biskra. The original equipment sank off the Arzeu beaches back at Oran. Ground transportation was a hand made motor scooter built by the engineering section using the wheels from an oxygen cart and a gas engine salvaged from an oil pump. An anti-aircraft gun was fashioned with a 20 mm cannon salvaged from a wrecked P-38, mounted on a fifty-five gallon drum filled with sand (what else) for ballast.

Living conditions at Biskra were primitive, there's no other word for it. Aside from the pilots in their five to a room Hotel Transatlantique it was strictly tents and foxholes beneath the date palms. Dining was al fresco making use of long tables and benches. Rank had no privileges when it came to food. One grabbed a mess kit, stood in line then sat or stood anywhere to eat.

The Germans soon recognized the concentration of aircraft at Biskra and began night raids usually by one or two Ju. 88s. However, in early January a large force of Ju. 88s really plastered the oasis. They hit both parked bombers and fighters and also squadron tent areas wounding several enlisted men. At the hotel the pilots, showing an instinct for going in the wrong direction, rushed to the roof to see the show. One bomb could have taken out most of them, fortunately nothing came close. Carl McCool,[4] one of the crew chiefs, says he was literally buried when a bomb hit the tent area and nearly filled his foxhole with him in it. Some of his buddies had him nearly dug out when they were all scattered by machine gun fire from a strafing pass or from ammunition set off by the bombing, hard to be certain which. Needless to say the foxholes got deeper after each raid. Glenn Hubbard recalled the foxhole depth question. Before the German raid some holes only went down six to twelve inches. After the raid some of

Bombed C-47 transport Biskra, December 1942. (Jack Ilfrey)

the holes really got deep. Too deep, of course, was just as bad as too shallow.

Anthony Sgroi, a 71st pilot, managed to get one for our side when he downed a raiding, or lost German bomber. He was sitting under his wing waiting for a mission call when the bomber dropped out of the clouds and went over the field at about 5,000 feet. Sgroi jumped in his P-38, gave it full power down the runway and then had to guess at the direction taken by the German. Guessing east he took out after it although it was well out of sight. After about ten minutes he saw a speck in the distance. The speck became a Ju. 88 flying about 500 feet below a cloud layer. Sgroi attacked and after several firing passes, the German crash landed in flames, still in Allied territory. The next day the crew was picked up and taken prisoner.

On January 20th during a fighter sweep in the Tripoli area the pilots of the 1st strafed truck traffic on the Gabes-Ben Gardane road. One pilot, Benton Miller of the 94th, had a direct encounter with a telephone pole. Somehow the plane stayed in the air and Miller nursed it back to Biskra. The plane should have won a prize as the most damaged yet flyable as the photo shows. One more added to the list of severely damaged planes and/or pilots. By January's end both were getting in very short supply and replacements were very slow in making an appearance. One source of aircraft for both the 1st and 14th Groups was the newly arrived 82nd Fighter Group. They were given the option of turning over half of their planes to the 1st and 14th at Youks les Bains and Biskra or turning over all their ships and going back to England and flying back more. The 82nd figured

Lt. Benton Miller's 94th Squadron P-38F-15, Ser. No. 43-2092, after tangling with a telephone pole gives testimony to the pilot's flying skill and the rugged nature of the Lockheed Lightning. (Ken Sumney)

that one ocean trip was enough and reluctantly gave up half their aircraft. By the time of the Casablanca conference in January 1943 the total strength of all three P-38 groups was down to ninety aircraft and drastic action was indicated. Additional P-38s were to be delivered as deck loads on ships from the USA as well as flown to North Africa via the South Atlantic, Ascension Island route.

Late in January the 71st was alerted for its move north to Ain M'Lilla. This would be accomplished on January 23rd where the squadron would be located in a grove of pine trees a good mile from the flight line. This fairly brief excursion by the 71st would be the last time that one of the squadrons would be located separately. Ain M'Lilla was about 90 miles north, not far from Constantine and considerably colder than Biskra. On the 22nd Lieutenants Landers and Dunn, two of the 71st's new pilots, were lost strafing vehicles on the Ben Gardane road. The move of the 71st signaled the end of the 1st stay at Biskra. The remaining squadrons would move north shortly bringing to an end an exhausting month which saw the unit dangerously low in both pilots and aircraft.

The 71st ended the month of January with a sort of double dip mission. Ten aircraft led by Captain Eiland, flew to Telergma for briefing and rendezvous with B-26s. The 71st pilots escorted the Maurauders to Sfax and were jumped by FW 190's near Kairouan airport some distance west of Sousse. This evolved into a running battle over the bombers as they headed west on the retirement route. The enemy aircraft persisted in making diving passes, pulling away when the P-38s turned into them. It was an inconclusive engagement with one P-38 and one Focke Wulf sustaining damage.

About this time the fifty mission cap was put in place together with the concept of rest camps. Both were timely. Fifty missions became a realistic but far from easy target.

The 5th Bomb Wing Commander gave a talk to the assembled officers of the 97th and 301st Bomb Groups and the 1st Fighter Group during which he announced a policy of rotation of the bomber crews at the completion of twenty-five missions. Amid uproarious applause he continued, "and you fighter pilots who have been flying alongside the bombers, you too can go home after twenty-five missions." There was an immediate flap when, after a quick check of the records, the staff realized he had almost freed nearly the entire 1st Fighter Group. The fighter pilot policy was immediately revised to read fifty missions but the goal had quietly changed. It was no longer defeat of the Luftwaffe, it became, "Do the fifty," instead.

Elmer Hartman, a Petdog pilot, described his first rest camp stay at Agadir, French Morocco:

> "Boy what a swanky place. Best location we've had since leaving the States. Hotel is situated on the side of a steep hill and the ocean is directly under it. Most of the rooms are single and have marvelous furnishings. No drinks and no women, so I suppose this will be as it was intended, a real rest. Certainly a grand night out, full moon shining on the ocean with the breakers coming in. Even have hot water and private showers for each room."

Agadir is located well down the coast of Morocco some 250 miles below Casablanca. The highly popular stay at rest camp was usually a week, in this case spent swimming, sunning, playing volleyball, etc. or just plain loafing. However, at the end of the week it was back to Biskra and the war in North Africa.

The scope of operations at this point in the war made it necessary to give recognition to the shortage of aircraft. On February 4th a sort of double effort was carried out by eighteen B-17s of the 97th Bomb Group

A 94th Sq. P-38F-1 destroyed in a Ju. 88 raid, Biskra. (Jack Ilfrey)

escorted by ten P-38s. This was followed by twenty-four B-17s of the 301st BG escorted by an additional ten P-38s. Compared to the massive escort missions later in 1944 these numbers seem exceedingly small. Another section hit an airfield west of Gabes and met much opposition. Captain Rimke claiming an Me.109.

Shortly after the Gabes raid, B-17s from the 97th and the 301st plus B-26s from the 17th BG were escorted to the Cagliari-Elmas airfield complex on Sardinia. This was the first Twelfth Air Force raid on a European target and Captain Newell Roberts claimed an Me.109.

The constant complaint in the 1st Fighter Group was, "When do we get more planes!" The old F models were nearly gone and the G's were showing definite wear and tear. This was evidenced by the number of early returns. Elmer Hartman mentions one mission with six returns out of fourteen planes due to faulty radios and/or engine troubles. He was one of the returns. However, replacement pilots were beginning to appear in many cases bringing additional P-38s. In at least one instance the replacements were asked to return to England and make a second trip bringing more planes.

Fifteen pilots, sent from the U.S. West Coast to England in January 1943, flew fifteen Lightnings from England to Algiers via Gibraltar with bomber guides. After one attempt to deliver the planes to the group in Biskra was turned back to Algiers by weather, Colonel Stone, went to Algiers and said he had pilots but not enough airplanes. He urged the pilots to return to England and bring back more P-38s. Departing on February 14th they did acquire more aircraft at the Lockheed installation in Belfast, flew to Southwest England and then to Casablanca. In mid-March these pilots were assigned to the 1st and joined the group at Chateau d'un. They included Lieutenants William Whelan, Dee Johnson, Max Rayburn, Robert Holcombe, Jim Hagenback and Archie Jackson assigned to the 94th; Frederic Kohn, Robin Hansen and Arthur Franke to the 71st and Alden Freng and Robert Anderson to the 27th. It should be remarked that the replacement pilots had, on the average, much less experience in the P-38. Jim Hagenback had 35 hours when he joined the 94th and was considered trained, some of the "new guys" would have even less.

Robin Hansen describes the England to Africa flight in the words of an inexperienced pilot. The first group of pilots to make the trip had more P-38 hours than anyone else on earth. Hansen had forty hours in the P-38 and little more than 200 hours overall which was typical for the new replacement pilots.

"Of the fifteen or so of us who started out, at least four were lost on that flight. We were loaded down with two 300 gallon belly tanks and staggered off the runway. B-26s were to provide navigation. Radio silence was imposed because the Germans were patrolling the Bay of Biscay. There was an overcast ceiling of several thousand feet. About a half hour or so out of Portreath an undercast materialized. Not too many minutes later the undercast and overcast merged and the B-26s went on instruments. You have to realize that instrument flying in that era was haphazard at best. Needle, ball and airspeed. Don't trust the artificial horizon. The formation being flown was not exactly tight, probably half because of inexperience, the other half because of the need to conserve fuel. At any rate most of the flight, including me, lost contact with the bombers. I remember the cardinal rule of instrument flying in those days. 'If you don't know where you are never go down, go up.' I climbed to get over the overcast and broke out a few minutes later. We must have been somewhere in the Atlantic but you couldn't prove it by me. I don't think we had much of a briefing on the headings to take. If we did I don't recall it. I was just following the B-26s. When I broke out on top there wasn't another aircraft in sight. I thought I'd had it and I hadn't even gotten to combat yet. I set course due east, but discovered after a minute or so, that I was following the adjustable course indicator on the flux gate compass and not the compass itself. I was turning to the proper direction when suddenly one, then two then three P-38s popped up out of the overcast. They were a welcome sight. A few minutes later as we were joining up, there was a break in the overcast and I spotted a B-26. Two of us dove down and this time closed in tight. The remaining P-38s set their own course and stayed on top.

Awards Ceremony - Biskra, 94th Sqd. R - L Murdock, Starbuck, Pringle, Malzahn, Ilfrey, Neale, Harman, Roberts. Presentation by Gen. Atkinson assisted by Col. Garman. (Lenhart)

"In another hour we emerged into beautiful

sunshine. We were a few miles off the northern coast of Spain or Portugal. The B-26 maintained 500 feet or so and stayed a distance from land. We began to see fishing boats. By this time most of my courage had returned and I left the B-26 and went down and buzzed several of the ships. When I was on the deck I noticed that my altimeter was reading almost a thousand feet although it had been set to runway elevation at Portreath. The drastic change in weather systems caused the altimeter to read high. I think that those who were lost on the flight chose to get under the clouds and not above them trusting their altimeter to avoid the water.

"We continued down the coast of Portugal through the Straits of Gibraltar landing at Gibraltar more than eight hours after take-off.When we got on the ground I found that my companion was 'Sully' Sullivan.

"We spent the night at Gib and flew to Algiers the next morning. The next day we got to Telergma which was another experience all its own. Telergma was a B-26 base and also home to the 82nd Fighter Group. As I pitched out to land, I could see that the B-26 just touching down was firing red flares for wounded on board. I landed and the B-26 which was to land after me burst into flames at the end of the runway. This was the real thing! That night I got together with classmates and drank wine in the makeshift club which was lit by candles.

"From Telergma we delivered our planes to the 1st at Biskra. P-38 and B-17 missions were being flown. I recall seeing a B-17 with the tail almost severed coming in for a landing with the fuselage collapsing part way down the runway. During our short stay there the French had caught several Arabs preparing to bomb our hotel. The trial did not last long. They were sentenced to be shot and taken by van with barred windows to sand dunes a mile or so from town. Most of us walked out to watch. I had seen scenes of this in the movies, but I was not prepared to see the Arabs looking out through the bars knowing this was their last day on earth. I would not watch again.

Operations Biskra L - R Neale, Murdock, Ilfrey, seated. (Jack Ilfrey)

"A C-47 took us to Ain M'Lilla to join the 71st. When I arrived a mission was returning from strafing near Gabes and a number of planes were missing. I believe that I was the first replacement that had been assigned. The remainder of the squadron had been together since before Pearl Harbor. I was not given the warmest welcome that I have ever received. In fact I was ignored. There were no provisions for billeting us. I was assigned to a sleeping bag vacated that morning by one of the pilots who had not returned from the days mission.

"A poker game was in progress in the tent with the players extolling the virtues of the missing pilot while ignoring my presence. I was wakened in the dark the next morning and asked if I wanted to go on a mission. This is an offer that is difficult to refuse. Off I went with a flight of four to Telergma where we were to be briefed on a mission. I think we were to be a part of the 82nd but I'm not sure. For some reason it was called off and I returned to Ain M'Lilla where I discovered that the missing pilot had returned and I had no sleeping bag. I also discovered that the missing pilot had fewer virtues than those he had been credited with the previous night."

Thus were Robin Hansen and the other raw replacements introduced into what probably seemed like a nearly closed society of the original squadron members. Nonetheless, replacement pilots as well as replacement P-38s would become more and more of a routine as the war took its toll.

Flying from and even existing at Biskra was becoming increasingly difficult. The dust and flying sand sometimes limited visibility to a few feet. The so-called Siroccan winds off the Sahara, increasing at this time of year, made operations extremely difficult. Sand, grit and dust penetrated everything and wreaked havoc with visibility, fuel lines, machine gun mechanisms, food, in short it was everywhere. Only camels seemed to operate effectively under such conditions. Herb Green recalls getting a scoop of canned cherries in his mess kit for dessert. By the time he got to dessert the cherries were unrecognizable they were so dusted with sand. Clyde Gabler said the only way to breathe was to put a wet towel over your head and hole up in a tent or bunker. A move to a new location was imminent.

Beginning about February 8th, 1943 the 27th and 94th squadrons began their move out of Biskra north to Chateau d'un du Rhumel. This latter was a small fighter strip about 125 miles north of Biskra not too far

"Big Friend", a Boeing B-17F, four-engine, Flying Fortress, on an African airfield. They were often escorted by the 1st Fighter Group. (Ken Sumney)

from the city of Constantine. The airfield was located on what used to be a race track. It would, in the spring, be a sea of poppies. Once again it was back in the mountains. While this provided a change from the sand, grit, date palms and desert Arabs of Biskra it was still February, cold and on occasion very wet in the mountains. Aside from getting away from the dust at Biskra the new location was considerably closer to targets in Tunisia, Sicily, Sardinia and ultimately Italy.

Early in February on an escort mission to Sfax two of the 82nd Group pilots flying with the 1st were hit. Lieutenant Drompp was shot down and killed; Lieutenant R. Hattendorf was hit by shell fragments. Hattendorf returned and continued flying with the 71st. There were two Hattendorfs on loan from the 82nd, brothers Wilbur and Richard, evidently part of a brief pilot exchange between the two units.

The months of February and March were difficult for the older pilots. Joel Owens says, "We were all experiencing some form of stress. Combat fatigue, colds, lousy living conditions and rumors. The news that the 14th Group had ceased operations had a tremendous negative impact on morale and some of the replacement pilots were depressingly inexperienced, many with little or no fighter time. One replacement pilot had only a total of twenty-five hours as a B-25 co-pilot, this since graduating from flying school."

Cold weather and tent living at Chateau d'un brought forth the design of a stove which, with few modifications, would be constructed and used in winter weather till the end of the war and in time it became G.I. issue. The body of the stove was generally a fifty-five gallon drum, or something of similar size, cut in half and upended over a piece of armor plate and a few rocks. A small hinged door was cut in the front. Fuel was aviation gasoline contained in an elevated tank outside the tent. The gas was piped to the stove via copper tubing, dripped onto the rocks by means of a control valve and lit. The chimney or flue consisted of tin food cans, the bottoms and tops cut off and hammered together. Lighting the stove by striking a match and tossing it in was somewhat chancey. The gas tended to explode if too much accumulated, singeing hands and scaring the devil out of the lighter and anyone else close by. Stuart Froehling, a replacement pilot who joined the 94th at Biskra, said it tended to jump up a foot or so. However, the heat was more than worth the risk as was the hot water and the occasional bit of hot food which could be cooked. Eggs, when they could be had, and grilled cheese sandwiches were particular favorites. Stuart Froehling also gave favorable mention to the British liquor ration available early in 1943. One fifth of whiskey and one of gin every two weeks. Stuart kept the whiskey and gave the gin to his crew chief (rank had its privileges).

In mid-February the North African Air Forces were reorganized. On February 17th Brigadier General James Doolittle took command of what was designated the North African Strategic Air Force. The NASAF included the XII Bomber Command and the 1st and 82nd Fighter Groups. (The 82nd had made the England-Gibraltar-Africa flight in December 1942). Also in mid-February Captains Owens (27th), Roberts (94th), Eiland (71st) and Lieutenant Ilfrey (94th) received a commendation from General George Marshall via the Twelfth Air Force. These four were included among the 15 leading fighter pilots in the North African theater.

On February 21st everything nearly came apart. Under cover of extremely bad weather the Afrika Korps attacked through the Kasserine Pass in Southwestern Tunisia, badly mauling the American 1st Infantry Division. On the 23rd with clearing weather the 27th and

94th strafed through Kasserine and encountered intense anti-aircraft fire. Jack Ilfrey says the Germans were firing up, level and even down at the Lightnings. Of twelve making the strafing run only eight made it back to base. Two went down in the Pass (Rimke and Chambers) and two more crash landed behind Allied lines north of Kasserine. Nearly every plane was hit. Captain Newell Roberts, who led this mission describes it in the following terms.

"The weather was extremely bad and it was by no means clearing. Orders came down from General Doolittle to Lieutenant Colonel Garman to take thirty-two P-38s and strafe German troops and vehicles in Kasserine Pass. The message was received late in the afternoon and some three hours after our troops really needed support from the Air Corps. I was asked by Colonel Garman to get the flight together and to lead the mission. I stated that I would, I knew the territory and had been there many times before. I could only get sixteen P-38s that were flyable in the entire group. It took 45 minutes to brief the pilots and get everyone airborne, then of the sixteen ships available only twelve finally made up the formation due to maintenance problems.

"It was a very dangerous mission and I told Captain Rimke that he did not have to go. He was my closest buddy, however, he insisted on going. We decided to attack Kasserine line abreast with Captain Rimke and his wingman on my left and Jack Ilfrey on my right. We flew about 50 to 100 feet altitude below the solid overcast. The overcast was not more than 200 to 300 feet. As we flew over our own troops I have never seen so many dead and dying men. There was no place you could have stepped immediately before entering the Pass without stepping on a body. Just as we got over our own troops they opened fire at us with everything they had. Practically all our P-38s were hit with our own fire. Rimke's ship was hit badly and he was killed. The Germans were met head-on firing at us. We made one sweep into the Pass, then pulled up into the overcast on instruments, avoiding the mountains and made a gentle climbing 180 to the left flying for some time then letting down through the overcast and back to base at Chateau d'un."

The next day the attack was repeated mustering all the P-38s we could get in the air. Harold Lentz of the 94th describes this mission:

"The flight plan was to fly in group formation to the south end of the Pass. If the Pass is clear we will make a fast run through hitting targets of opportunity. North of the Pass we will fly around for half an hour, make a return run through, and come home. Squadrons were to enter the Pass with flights at trail. They would then reform in the normal squadron and group formation to avoid overlapping any weapon trajectory. There was a second pass twenty miles to the west. It may be available to pass through to the north if the weather at Kasserine is marginal. It may be manned. In this case make a one time run through, go to the north end of Kasserine and make a one time run, north to south through the Pass (Kasserine). Then come home.

"The sky south of the Pass was clear. There was a thin inversion layer of clouds over the area of the Pass but sunshine was evident in the pass itself. Cockney Red leader elected to proceed on through the Pass. He made a wide 360 degree swing to string out the squadrons and the flights and headed out to enter the Pass. Squadrons and flights fell into line as planned with radio silence in effect.

"The scene within the pass was pastoral. The road wound down among hummocks and large rock formations then flattened out to wend its way north across a broad grassy plain. Sunlight was filtering into the Pass through an opening between the cloud cover and the west wall of the Pass. About a quarter of the way into the Pass an American GI truck lay askew on a boulder; the only evidence of a battle.

" Nothing within the canyon was identifiable as weaponry or tactical military equipment. There was an obscure line of what appeared to be hummocks, boulders and/or structures strung across a narrow point at the north end of the plain which made up the middle half of the Pass. The group was more than a third of the way into the Pass and nothing had happened.

"The squadrons were in left echelon position to the lead squadron but held the normal blue, red, white pattern within each squadron. All were on the deck and holding in steady form, showing no evidence of anxiety, a beautiful sight.

"About halfway into the Pass it became evident that the obscure line was military vehicles. One could have been mistakenly identified as an ambulance because of what appeared to be a long stemmed Red Cross on the side. It wasn't the type used on U.S. vehicles. Such a display seemed to be contrary to a military unit's security. The thought that crossed my mind was -'Boy, what a shooting gallery!' I became concerned if this was a cross fire entrapment. At this point the words, "Stop Shooting Fellas It's Our Own Stuff", came on the radio.

"No one was shooting! The vehicle line wasn't in range. Cockney Red leader pulled up sharply and my concern was now elevated to speculating on another Rommel ruse. Did Rommel have a radio with our frequency through which this call was made?'

"In the next instant Cockney White 3 was hit. He took his ship up through the clouds and a few minutes later came diving back down. The pilot jumped off the trailing edge of the wing and cleared the tail before cracking the chute.

"When I looked back Red flight was gone. Blue flight rolled to the left to head for the canyon floor. I had seen a curtain of high explosive shells bursting just beyond the line of vehicles that strafing aircraft would have to fly through while lifting over the vehicles. Yes, Rommel's ruse was working and we were the turkeys in his turkey shoot.

"I used a cross control skid to bring the flight down, thus throwing off the marksmanship of the gunners. I noted P-38s going in several directions but couldn't spot any targets even though I ran a line of trajectory of a 38 that seemed to be zeroing in on something. Nothing could be seen. The mission and the group was in a disorganized shambles. With distance running out I took the flight up over the west wall of the Pass to seek out the alternate Pass mentioned in the briefing.

"When we broke over the wall we found ourselves in an east to west valley overlaid with the same cloud cover overlaying Kasserine. The valley closed in and down to the west. The clouds closed with it to a dark area at the west end of the valley. This left maneuvering room for no more than one aircraft in a straight down dive-bomb run or Immelman turn off a strafing run. There was no evidence we were in the right Pass. I turned the flight to the southeast and began to climb up toward the overcast. As we leveled out I looked back. A five-barreled AA gun was firing. The Pass was indeed manned. The guns were secreted back into the valley we had entered, a devious entrapment for anyone using the alternate Pass."

Fortunately, clearing weather as well as reinforcements stiffened resistance on the ground had convinced Rommel (actually Von Arnim) to withdraw back through Kasserine. In the process, however, the American infantry learned a costly, bloody lesson as did the Air Corps. The Afrika Korps may have been retreating but it remained an extremely dangerous adversary, particularly the panzers. Had the Kasserine breakthrough succeeded, the Africa Korps could have reached the Mediterranean, splitting the Allied Armies and generally wreaking havoc.

Late in February 1943 the 71st squadron moved to Chateau d'un to join the 27th and 94th completing their last wartime separation.

Beginning in March and extending into April the makeup of the squadrons began to change. With a more or less steady flow of replacement pilots the veteran cadres of all three squadrons began to leave. From the 71st Major Rudell transferred to XII Bomber Command and Francis Harris, Anthony Sgroi, Luther Morefield, Raymond Parsons, William Morhaus, Bernard Muldoon, William Neumann and Vincent Rethman, departed. Captain John Eiland took over as the 71st commander. In early April Newell Roberts, Lewis Murdock, Jack Ilfrey, Robert Neale, Paul Forsythe, Clifford Molzahn, Donald Starbuck, Earl Hille, and Wesley Pringle of the 94th headed for home. Glenn Hubbard was posted to the North West Africa Strategic Air Force and James Harman replaced him as Commanding Officer of the 94th (on March 6, 1943). From the 27th, Harold Mendenall, Elmer Hartman, Marcus Linn, Marvis Morrison, Joel Owens, Roger Stemen, John Wolford and Robert McCord all departed. Major Weltman was transferred to group and Captain Jim Pate assumed command of the 27th Squadron.

For most if not all of these men departure was not an immediate ticket back to the States. The Air Corps seemed very uncertain about what to do with them. Many were assigned to operational training units in North Africa, that is, training new pilots in tactics based on the considerable experience of the veterans. More often than not however, they wound up assigned to the wrong planes, Elmer Hartman and several other pilots in the same status were assigned to train people in P-39s or P-40s. All of which made little sense to the pilots.

A separate group of experienced pilots was transferred to the Training Command for assignment to the 14th Fighter Group. These pilots included Major Joel Owens, Lieutenants John Wolford, Robert McCord, Robert Riley and Captain Roger Stemen from the 27th, Captains Bernard Muldoon and Vern Schoenfeldt, Lieutenants William Newman and Vincent Rethman from the 71st and Joseph Smith from the 94th.

A B-26 Marauder, frequently escorted by the 1st Fighter Group, crashes at the 1st Group's base with battle damage and breaks its back. Unit unk. (Sally Sweet Stangl)

1st Ftr. Grp. pilots involved in reactivating 14th Ftr.Grp. Standing L - R Rethman (71st), Stemen (27th), Owens (27th), Capt. Eddie Rickenbacker, Muldoon (71st), Wolford (27th), Smith (94th). Front L - R Newman (71st), McCord (27th), Riley (27th), North Africa, May 1943. (Rethman)

Joel Owens describes his new assignment:

"I was relieved from assignment to the 27th at the end of March. I was a Major who had flown seventy-four missions, scored seven aerial victories (5 destroyed, 1 probable, 1 damaged). Had been recommended for the Silver Star, the DFC with four oak leaves plus a bunch of Air Medals. The squadron had fared pretty well in combat. The record stood a sixteen losses versus sixty-four victories (35-7-22) all but three were single engine fighters. I was really looking forward to stateside duty.

"I hadn't attached much significance to my orders which directed me to report to the Northwest African Air Force Headquarters. I arrived at Casablanca, checked in to transient quarters and headed for the Officers Club. I ran into several pilots from the 1st Fighter Group. We were having a rather boisterous time celebrating our escape when Major Joe Ruble came along. Joe had been a member of the 1st Group staff in England. After an exchange of pleasantries he said, 'Don't drink too much tonight. General Cannon wants to see all of you in his office tomorrow morning.' Nobody had any idea what lay ahead. We all agreed that he would probably hand out some medals, tell us what good boys we had been, and wish us well.

"The next morning some of us were a bit bleary eyed as the General arrived. He was almost fatherly as he congratulated us on our records. Then he asked: 'What, in your opinion, is the one thing most needed in this Theater of Operations? We all agreed that what was needed was MORE FIGHTERS!! The General reacted with: 'I was hoping you would say that. Now here is what we are going to do. We are going to resurrect the 14th Fighter Group and you men are going to be its leaders. I don't remember that any of us thanked the General for this splendid opportunity.

"The next day we reported to Lieutenant Colonel Troy Keith, the only rated officer from the original 14th Group. He spoke briefly with the dozen or so pilots who had been ordered to Mediouna, a small village southeast of Casablanca. Colonel Keith told us that the 14th Group finally consisted of three squadrons. The 48th and 49th with ground crews and officers but no pilots or aircraft, and the 37th which had a complete complement of personnel but no aircraft. The 37th was commanded by Major John Bright, an ex Navy pilot and Flying Tiger ace. No one seemed to know how they came to be in North Africa since the rest of the 37th's former 55th Fighter Group remained in England. Captain Carl Rudder, ex 94th and victim of the Greenland ice cap escapade, was a member of the 37th as was Marcus Linn. Linn had joined the 27th in England from the RAF.

"I was the Group Executive Officer, Muldoon and Newman were appointed as Squadron Commanders of the 48th and 49th squadrons. The rest of the 71st pilots joined the 48th and the 27th men went to the 49th. They were assigned as flight leaders, operations officers and such. We began the task of bringing the group to operational readiness. The 14th had a high priority on the acquisition of both aircraft and pilots. The 78th

Fighter Group in England was converting from the P-38 to P-47's and started ferrying P-38s to Casablanca. The first group of them delivered their planes and immediately hopped a C-54 back to England. A quick call to Major Ruble resulted in the reassignment of the pilots on subsequent flights. Some of them were unhappy about being Shanghied, but we explained to them that 'War is indeed HELL'. The group was at full strength in about ten days. After training in formation flying, gunnery, bombing and high altitude combat maneuvers the 14th moved to a forward base southeast of Telergma on May 5th and flew its first mission on May 6, 1943. This, to my knowledge, is the only instance of a new group entering combat with leaders who had combat experience. It was a pretty good combination. The Old Heads had the knowledge of how to avert disaster and the new guys were fearless and full of vinegar or whatever it takes to make a fighter."

Most of the 1st Fighter Group pilots assigned to the 14th were returned to the States by late June. They had lost one; John Wolford an ace with the 27th, was KIA on May 19th. On June 1st Major Joel Owens flew his 87th and final mission with the 49th squadron before returning to the USA. The part played by Major Owens and his fellow pilots in "restarting" the 14th Fighter Group completed the role of the 1st Fighter Group as "father" of the P-38 units in the Mediterranean Theater. In early 1942 in California the 1st had contributed to the fledgling 82nd Group, senior officers, experienced pilots and higher ranking NCO's including people capable of stepping in as line chiefs. Both the 82nd and the 14th Fighter Groups would complete the war as near neighbors of the 1st in Africa and in Italy. In fact we would have to move in with the 82nd for a couple of months in the winter of 1944-45. Probably our last contribution to the other groups was Lieutenant Colonel John Weltman who moved from 1st Group Executive Officer to 82nd Group Commander in early May 1943.

Returning to the tribulations of 1st Fighter, Glenn Hubbard recalled starting inexperienced replacements in with orientation flights with the experienced pilots. After a time Hubbard told the new men they would start flying with the experienced pilots on what they thought would be relatively tame missions. On the first so-called tame mission the squadron was jumped by a large cluster of Me.109s and had to scramble for their lives. The new pilots were stunned by what they had been told would be a "milkrun". .

Two of Archie Jackson's tentmates at Chateau d'un were Clark Jennings and Benton Miller, both of whom had been shot down and returned. Jennings had come down in Allied territory, had not been captured and was allowed to finish his missions. Miller had had a somewhat different experience. After being shot down he was picked up and held by a group of Arabs. While debating what to do with him one of them took his Mae West (an inflatable life jacket) and while looking it over apparently triggered one of the cylinders. The vest inflated scaring the Arabs who tumbled out of the tent. However, they then crept back and attacked the vest, knifing it to ribbons. This in turn scared Miller who thought he might be next. The Arabs apparently had the idea to auction him off to the highest bidder amongst the French, English, Germans and Americans. However, he was finally bartered to the French who turned him over to the English who eventually returned him to the 94th. Although Miller had been captured and escaped after a fashion, he was allowed to continue to fly missions. There was a fairly general policy stating if one had been captured by the Germans and escaped, one's flying days in that theater were over, however, Miller had not been captured in the normal sense.

Names on this 94th P-38E, Ser. No. 41-2053, include Unice, Ace, Daisy and for whatever reason, The Rodent. Chateau d'un. Pilot and Crew Chief unknown. (Sumney)

94th Sqd. pilots Chateau d'un, April 1943. Back L - R Stover, Fell, Hagenback, Whalen, Holcombe, Stephen, Brophy, Miller, Froehling, Wellensiek, Burger, Sutcliffe. Front L - R Rayburn, Newsome, Johnson, Jackson, Jennings, Brown, Quinn, Harman, Reese, Boose, Lentz. (Jackson)

During March the group began concentrating on escort missions, ranging farther afield and with an increasingly stronger bomber force. On March 3rd Harry Dowd was killed on an escort mission to El Aouina. March 8th was a better day. All three squadrons escorted B-17s on an anti-shipping mission east of Bizerte. The Germans were now doing their best to reinforce their hardpressed troops. As the P-38s and bombers reached the convoy the Lightnings attacked the German escort of Me.109s and FW 190s. In a swirling dogfight we destroyed five 109s and two Focke Wulfs; probably destroyed an additional five Me.109s and damaged two190s. We lost none although we did suffer some machine gun hits and at least one engine shot out. Harold Lentz was shot down by a P-38 apparently flown by a German pilot in late March 1943. He crash landed in a desert canyon in North Africa after being well shot up by an unmarked Lightning. Lentz gives a compelling description of a near panic escape from a burning ship. He came to draped over the control column.

L - R Capt. Jim Harman, Lt. Gen. Carl Spaatz, Lt. Col. Ralph Garman, Brig. Gen. J.H. Atkinson. Harman, CO 94th Sq., was killed in action shortly after this photo was taken. (Lenhart)

> "I reached for the canopy release, found it jammed; reached for the right window handle, found it missing; reached for left window handle and the window was jammed. I thought to put back to canopy for leverage, but my left foot was stuck between control column and rudder. Told myself to unzip boot-had Arabic slipper on under boot. Unzipped boot, freed foot, tried to stand up in cockpit but my safety belt still hooked. Sat down, unhooked belt, started to get up again, parachute still on; sat down, thought I had unsnapped chute, but hadn't. Talked to myself and carefully unsnapped each buckle on chute. Put back against canopy but discovered that had punched hole in same evidently with hard German head. Worked through hole in canopy only to discover that throat mike still fastened. Also I was dripping blood from cut eyebrow and hand. Realized engine still running and fire in radio compartment."

Lentz crawled out to find a group of Arabs standing around and the unknown P-38 circling. He arranged to destroy the ship and got the Arabs away from it. They fortunately were friendly, put him on a horse and took him to a town where they turned him over to a French colonial type in exchange for a bolt of cloth and a pair of shoes. From there, with the aid of the American Army, Lentz made his way to Tebessa and then back to the outfit. He continued to fly missions and returned to the USA at the end of July 1943.

Early in March bomber command was augmented by the arrival of the 99th and 2nd Bomb Groups each flying B-17s. In early April a total of 91 B-17s attacked the Naples-Capodichino airfield area, the first time Naples had been attacked by the Twelfth Air Force.

As a matter of information there were B-24's in North Africa but not in the Twelfth Air Force which had

Red Cross workers Eleanor Preble (left) and Sue Macpherson provide coffee and donuts to appreciative 27th Sqd. pilots: L - R Radcliffe, Kennedy, Welch, Bancroft, Stevens, Loder (kneeling), Pate, Ross, Newby, Szaflarski, Shaw Brown. (Szaflarski)

B-17s as the only heavy bomber. The B-24s at the time were assigned to the Ninth Air Force and were operating far to the east out of Libya. Later in the year these B-24's would take part in the massive low level raid on Ploesti. The first of many costly trips to this oil refinery complex.

One item deserving mention appeared in the 71st squadron minutes at this juncture: "Well, well,well - will wonders never cease! Today we were served coffee and doughnuts. That's a wonder in itself but the big surprise came when none other than American Red Cross girls did the serving—they are going to visit us twice a week." This comment appeared in mid-March.

April 1943 saw the beginning of the end in North Africa. Operation FLAX began on April 5th concentrating on the destruction of German and Italian air transportation into and out of Tunisia. The water transport system was targeted as well. Germany and Italy were using large scale air transport to ship reinforcements into Tunisia. Allied forces did their best to interfere. Probably the best known example of activity in Operation FLAX was the so-called "Palm Sunday Massacre". On that day Allied aircraft decimated a large formation of Ju. 52s and Me. 323s, the latter huge six engine flying gliders. The 1st did not take part in the Palm Sunday affair but was involved earlier in April in massive fighter sweeps of the Sicilian Straits. On April 5, with the 71st providing top cover the 27th squadron and the 82nd Group jumped an enemy formation consisting of fifty to seventy-five Ju. 52s, six Stukas, twenty Me.109s and four FW 190s. The combined Lightning unit destroyed eleven Ju. 52s, two 109s and two Stukas. The 82nd downed an estimated fifteen of the Ju. 52s. The 71st Squadron got left out of the scoring but engaged the fighter escort to permit the others to attack the transports. Two pilots were lost, both from the 27th, Donald Field and Donald Hilgert.

The same sort of mission took place again on April 10th. Twenty SM 75 and 82 transports and eight Me.109s or Macchi 202s were destroyed without loss. This mission was led by Colonel Garman with Captain Eiland leading the 71st which flew the low echelon and accounted for most of the transports.

Junkers 52 transports of the type shown here, near Tripoli in 1941, were the aerial pack mules of the Africa Korps. (National Archives)

27th Squadron pilots, L - R) George L. Ross, George A. Rush and Sam F. Sweet discuss the April 10, 1943 mission in which each scored a kill. Flying top cover, the 27th engaged enemy fighters and each of the pilots downed an Italian Macchi 200. (Sally Sweet Stangl)

Lieutenant John Moutier, flight leader, responded to the call, "Transports at two o'clock," brought his flight around in a steep turn cutting in between the tri-motors and the (Macchi) fighters and began to bear on the target. About halfway around in the turn he saw the Macchi's coming in on his left wing and turned into them. Rivers, leading the second element, took a couple of shots at the fighters then pulled up on a transport. Rivers saw his tracers go across the cockpit and engines, the transport burst into flames and went into the sea. Walter Rivers pulled up to the left to make a 360 but he never completed it. About halfway around the turn a Macchi scooted past him. He turned onto the fighter's tail and shot it into the water. Completing his 360 turn he came around on another SM. A twenty degree deflection shot put this one in the ocean. Number four was dead ahead but close. He sprayed the fuselage and cockpit and then overshot. Beginning another 360 degree turn he closed almost head-on into another transport. This one went down in flames almost directly under him. Time to reform and head for home. Elapsed time, less than ten minutes.

Twenty-five ships went down that day, none of them ours. Bear in mind these were transports representing Rommel's attempt to fly in replacements for his Africa Korps. The P-38s left a trail of smears of smoke and bits of wreckage marking the funeral pyres of infantry replacements who had to sit in those tin coffins knowing there was nothing they could do but expect a machine gun bullet or a flaming dive into the sea.

On April 12, 1943 all three squadrons escorted B-17s on a mission attacking both shipping and the Tunis docks. A running dogfight broke out over Tunis resulting in the destruction of two Me.109s and two FW 190s. Meldrum Sears and Willard Bolton of the 71st each claimed a Messerschmitt; a probable 109 was shared between George Sutcliffe and Leonard Stephan of the 94th. Edwin 'Hank' Schneider and Lee Wiseman of the 71st each claimed an FW 190. In the debriefing following this action the pilots noted the presence of six Siebel Ferries in the Gulf of Tunis. These ferries were large motorized barges for troop transport. They were heavily armed, particularly with antiaircraft artillery and very difficult to hit.

The large losses inflicted on the Luftwaffe particularly during March and April severely hampered German air operations. Many of the Ju. 52 and Me. 323 pilots were instructor pilots drafted for the purpose of ferrying troops to Tunisia. Germany could ill afford to lose such people. Overall the Axis North African air losses were huge. Between November 1942 and May 1943 the Luftwaffe lost 2,422 aircraft in the Mediterranean theater.

Lieutenant Ed "Bo" Shaw, 27th Sq. poses by his P-38 in Spring 1943. He was downed and taken prisoner on 5 June 1943. (Sally Sweet Stangl)

With improving weather - excepting fairly frequent rains - ground sports activities began in earnest. These included volleyball leagues, baseball leagues and Monday night fights, which became Sunday afternoon fights. Volleyball tended to be intrasquadron, baseball included everybody with a team and the fights more often than not pitted our guys against local Arab fighters and some of the latter were very good indeed. Technical Sergeant Eugene Sarchione a 71st squadron boxer, fought the Arab champion of Constantine twice and was barely decisioned both times. Speaking of Arabs we should include mention of GI Joe an Arab man of all purpose more or less adopted by the 71st squadron. Joe worked around the kitchen and camp and was paid by the men chipping into a pool. One April payday Joe pocketed about 3,700 francs. Probably more money than he had ever seen in his life.

As time permitted between missions the 1st continued its efforts to develop dive-bombing techniques with the Lightning. Lieutenant Colonel Weltman and Captain David Stentz became the prime movers in this endeavor, something of a cross between a science and an art form. The pilots had been working on dive bombing techniques by practising on targets drawn in the desert. Little was known about dive bombing; some were concerned that the bomb, after release, might hit one of the propellers. As it turned out the P-38 was moving a lot faster than the bomb terminal velocity, but it did take some practice to get the hang of aiming the ship, when to release the bomb, i.e., low enough to be accurate but not so low that one flew into the bomb blast. What to do after releasing the bomb was generally stay low and get out of the area fast. Weltman would soon transfer to the 82nd Fighter Group, the 1st Fighter's final and significant contribution to the other P-38 units, but before he departed he left a lasting impression on enemy shipping in the harbor at Porto Torres, Sardinia. On a bombing mission on April 13th our P-38s scored a hit on the Cruiser Derizia. Weltman skipped his bomb on two freighters moored side by side at Porto Torres. The bomb went through the superstructure of the nearest freighter and exploded in the second, destroying and/or damaging both as well as a good bit of the surrounding dock area. An after the war review indicated that a good number of our delayed action fuses used in skip bombing were damaged during the drop and the bombs never exploded.

By mid-April the attention of the B-17s was shifting more and more to Europe. The attention of the 1st Fighter shifted also, presaging the invasion of Sicily. Palermo and particularly Palermo harbor became a frequent and, at times, a costly target. Escorting B-17s on April 17th Captain James Harman, the 94th C.O., downed a Ju. 88 but in turn was hit himself and forced to parachute. His chute was reportedly strafed by the enemy. He did not survive nor did Lieutenant Anderson of the 27th. On this same mission Lieutenants Matthews and Lowe claimed Ju. 88s and Lieutenant Harold Lentz probably destroyed both an Me. 109 and an Me. 210. Harman's tenure as Squadron C.O. was on the order of six weeks. He was replaced by Captain George Sutcliffe, the last of the "originals" with the 94th.

On April 13, 1943 the group escorted B-17s to Caglairi harbor. One of the bombers was downed. Unfortunately, an unofficial passenger on that particular bomber was Captain William Cody, 27th Squadron Intelligence Officer, who wanted to experience a combat mission. Cody was captured and became a prisoner. Rumor had it that Cody had in his possession about 2000 U.S. dollars so, he must have been one of the richest P.O.W.s ever taken. Colonel Garman was less than pleased at this misadventure.

In early May Eddie Rickenbacker again visited the group and the 94th to somewhat mixed reviews. The enlisted contingent stood in line for what seemed like hours only to be told that Captain Eddie would speak only to the pilots. The pilots were not impressed to be told how rough things were in the Pacific when they were losing nearly one out of two themselves. After touring Chateau d'un, Rickenbacker announced that he was throwing a party that night at a hotel in Constantine. Archie Jackson describes the affair.

"There were then some twenty-two pilots in the 94th squadron. They were trucked in to Constantine and a very nice meal indeed was laid on, camel steaks and the works, all washed down with cognac and grapefruit juice. Also attending were Generals Doolittle and Spaatz.

27th pilots at Chateau d'un. Top: L - R McKay, Pate, Newbury, Diesenroth, Brown, Nielson, McQuown, Hloucal, McIntosh. Middle: Smythe, Hurst, Stunkard, Wellner, Styer, Bancroft, Holden, Szarflarski, Butler. Bottom: Weinberg, Wenergren, Kuba, Shahan, Kennedy, Purvis (Caughlin)

After the meal Captain Eddie assembled all the pilots on a small stage; Doolittle called each pilots' name, Eddie presented a small, silver Hat-in-the-Ring pin and General Doolittle pinned it on the tunic above the pilots wings. General Spaatz then told us it was legal to wear this pin above the wings on his authority."[5]

Within weeks after the pin presentation nearly everyone in the squadron had a pin. They were nicely made out of silver coins, either a silver dollar, British half crown or silver half dollar.

Some time before the Rickenbacker visit the 94th had been adopted by a small burro or donkey. He ambled in and out of camp, usually at mealtime. Meals were taken out of doors-the usual mess line with stand up tables made of planks and stakes. There was no officers mess and no rank in the food line, and not much in the way of food either. Shortly after the Rickenbacker visit a crowd was standing munching noon C-rations when the burro wandered by. One of the fellows glanced at it, thought a minute and said, "Fellows I have a name for our burro friend. From now on it shall be called 'Eddie the ass'". The burro persisted as the very unofficial squadron mascot for the rest of the stay at Chateau d'un. His hoof fronts and ear tips were painted yellow to match the yellow paint used on the P-38 wingtips and tail booms of the 94th for identification.

Rickenbackers' visit pretty well coincided with the end of Operation TORCH. The campaign in North Africa was over to all intents and purposes and the preparations for the invasion of Sicily were well underway. The last German forces in North Africa surrendered on May 13, 1943. The 1st Fighter would continue operating out of Chateau d'un for another month and a half before moving to Mateur in Tunisia in June, 1943.

One of the specialized duties performed by our pilots at this stage of the war was weather recon. These long (usually) single or at most, two plane flights might cover Sardinia, Sicily, perhaps as far as southern Italy, then home over parts of North Africa. The flight weather reports often determined bomber targets and were a necessary chore until better techniques and more appropriate aircraft became available. When someone was lost on such a sortie, they simply disappeared, often without trace. Whether they were hit by enemy fire, got disoriented by weather or hit a mountain was usually never discovered.

There was one rather civilized interlude in the midst of the air war. The mayor of Chateau d'un invited Brigadier General J.H. Atkinson, 5th Bomb Wing, and some of his group commanders to an impromptu cocktail party at the town hall. Colonel Garman attended with some of his staff. The hostess for the affair was the Countess De Blaney the owner of an extensive vineyard near the town. Her husband was away serving with the Free French forces. The following Monday night the Countess arranged a party for 1st Fighter Group pilots at the local theater. Refreshments included a wine punch laced with sugar and fresh oranges together with cookies made from home ground grain. A real sacrifice on the part of our French hosts and a wonderful treat for all of those able to attend.

1. The 14th ceased operations on January 28, 1943, reduced to just seven aircraft. The planes were turned over to the 82nd Fighter Group; ground personnel were withdrawn to Berteaux, and some older pilots were returned to the USA. Many newer, pilots were also assigned to the 82nd and the 14th was reequipped and reactivated in May 1943. The demise of the 14th was due in part to the danger of low-level strafing attacks but also to the piecemeal commitment of fighter squadrons-or air power in general- often at the whim of various army commanders. This situation was finally rectified with the arrival of Generals Spaatz and Doolittle. In his autobiography, *I Could Never Be So Lucky Again*, General Doolittle recalled that at one point one of the ground company commanders had requested fighter cover for a Jeep going out to repair a telephone line. He didn't get it.
2. In *Fighters Over Tunisia* it is mentioned that Umphrey may have been shot down by an Me.109 pilot from II/JG 53 however this seems unlikely since neither Hubbard nor Roberts mentioned a 109 in the engagement, a factor they would scarcely overlook. It seems more likely that Umphrey was hit by return fire from the bomber that went unnoticed by the others.
3. By JG 51 according to *Fighters Over Tunisia*.

Egg trading Chateau d'un - 27th pilots, L - R McIntosh and Hloucal. (Szaflarski)

4. McCool and some others left the 1st in July 1943 transferred to the CBI theater.
5. As an aside the Hat-in-the-Ring was indeed the 94th emblem however, your author was never able to find written authorization to wear the pin above the wings. Quoting Captain Eddie didn't work too well back in the States with unbelievers who insisted that nothing could be worn above the wings. After his return to the States Harold Lentz did write to Rickenbacker about it, however, the latter was reluctant to involve General Arnold. Probably just as well.

1st Fighter Group Identification Markings and Call Signs

Effective September 10, 1942:

	Markings	Call Signs
27th Fighter Squadron	HV	Petdog
71st Fighter Squadron	LM	Cragmore
94th Fighter Squadron	UN	Springcap

Effective February, 1944:

27th Fighter Squadron	1 to 30
71st Fighter Squadron	31 to 59
94th Fighter Squadron	60 to 90

Notes:
1. The yellow TORCH circle around the national insignia was used at the time of the North African invasion, then removed, and then the blue bar and red outline were added for a time in 1943.
2. At various times additional call signs were intermittently employed. They were Cockney, Doorknob, Warrior and Highlife.
3. Squadron colors began to be applied in 1943 as follows:

27th Sq.	Red band on tail booms, wing tips and prop tips.
71st Sq.	White band on tail booms, wing tips and prop tips.
94th Sq.	Yellow band on tail booms, wing tips and prop tips.

4. When natural metal finish aircraft came into general use the 71st Squadron changed to black.
5. Red prop spinners were introduced in 1944.

A derelict Messerschmitt 109F of JG 53 in Northern Tunisia. (James V. Crow)

A pair of German Africa aces who fought against the 1st Fighter Group, Oberleutnant Franz Schiess (left) and Leutnant Hans Roehrig, both from III/JG 53. Schiess was eventually killed over Italy with 67 victories to his credit. Roehrig, who scored ten kills in Africa, had a total score of 75 when he was killed during the Sicily invasion. (James V. Crow)

An abandoned Italian Macchi 202 fighter in North Africa. The 202 and the Me. 109 were often mistaken for each other at a distance. (James V. Crow)

7

BLUE MEDITERRANEAN SKIES

Churchill called Southern Europe "Hitler's soft underbelly" and it turned out to be anything but descriptive. In the words of historian Samuel Eliot Morison, "The soft underbelly turned out to be stone. It proved to be boned by the Apennines, plated with the hard scales of Kesselring's armor and shadowed by the wings of the Luftwaffe." Allied land armies, the American Seventh and the British Eighth, would tackle this soft underbelly through Sicily and Italy and find it girdled with steel. Air attacks from North Africa most frequently focused on Sardinia and Sicily, less frequently on Italy. These attacks were in preparation for the invasion of Sicily still two months away. The raids on Sardinia served two purposes. It confronted the enemy with a choice of sorts-would we invade Sicily or Sardinia. The Germans probably didn't have much doubt that Sicily was next but Sardinia remained a possibility. Sicily is the largest island in the Mediterranean and has been a mountainous crossroads for thousands of years. Sardinia in contrast, while nearly as large was far from a crossroads. For centuries it had been a remote, malaria infested backwater. Our main objectives on Sardinia were the German and Italian airfields in the area of Cagliari and shipping in the nearby gulf. With the invasion of Sicily the bases in Sardinia had to be attacked to prevent raids on the invasion troops and ships. But first there was Pantelleria.

In May 1943, in addition to targets in Sicily and Sardinia, air attacks were launched against the island of Pantelleria, a small, rocky Italian possession located in the Mediterranean between North Africa and Sicily. The taking of Pantelleria was dubbed Operation CORKSCREW and this could prove no easy task for an invading force. The terrain of the island was hilly with numerous ravines. Hundreds of high stone walls dividing the arable land into fields, would provide formidable defensive positions and the enemy had positioned numerous pillboxes and machine gun nests in addition to the usual clusters of antiaircraft guns. The 1st was involved in the attacks on Pantelleria beginning early in June. Probably a half dozen missions were flown in quick succession just prior to the surrender of the island. Pantelleria was hit by just about every aerial weapon with the possible exception of torpedoes. B-17s and medium bombers made their runs and the P-38s added dive-bombing, strafing and skip bombing. Pantelleria surrendered on June 12, 1943 the first ground objective ever to surrender to air power alone. The fighter pilots were aided in their efforts against Pantelleria by a large photo mosaic of the island constructed by group S-2. According to the records the only Allied casualty in the occupation of this island was one soldier bitten by a jackass. Presumably he recovered but his dignity must have been severely wounded.

In the intervals between missions there was little to do at Chateau d'un in the way of recreation. Once in a while people would drift down to a little nearby creek and get in some 45 pistol target practice shooting at snapping turtles. This was infrequent because the powers that be kept a close rein on 45 caliber ammunition. One of the local sheiks, in a rare display of friendship, invited some of the group to his "castle" and put on quite a feast complete with roast goat and good wine. This was unusual because relations with the Arabs were just not that good. The natives were a good source of eggs (100 eggs for 200 francs or about $2 American) and eggs were about the only good thing to eat, a break from the Army chow which seemed to specialize in corned beef, Spam, beans and stew and more beans and more stew. Unfortunately some of the Arabs tended to steal anything that wasn't tied down, even to the extent of crawling into tents at night. And they put stolen

94th P-38's at Mateur. Not all ships had romantic names. Right foreground says FLYIN' BUCKET A'BOLTS, a G-10 Model. (Sumney)

June 1943, Lieutenant Dan Kennedy's 27th Sq. plane BEANTOWN BOYS is about to get a 500-pounder slung under the wing. Kennedy shot down three Me. 109s over Pantelleria on 9 June to become an ace. (USAF)

goods to strange uses. A barracks bag with two holes cut in the bottom and cinched in at the waist made a serviceable pair of pantaloons. The Arabs had other goods also. Archie Jackson and his tentmates bartered for rugs to give their tent a floor of sorts. They got the floor and an infestation of fleas to go with it. How to get rid of the fleas? Roll up the tent flaps and douse the rugs with aviation gas, and for God's sake don't light a cigarette! By nightfall both the fleas and the gas were gone.

There were occasional visits to Constantine, an old French-Arab city built on two high cliffs connected by suspension bridges. Not much to do in Constantine other than sightsee and an occasional movie, although there was a casino or two featuring assorted bad liquors and girls of about the same quality with stage shows and the like. Nonetheless, Constantine did have a touch of the exotic-most French cities do-and was certainly more interesting to visit than our tent villages. The city seemed to have survived its German and Italian occupiers none the worse for wear although some of the native quarters couldn't be too much worse.

Missions in early May 1943 nearly alternated between Sardinia and Sicily with the latter tending to get more the brunt of things as the month wore on. Escorting B-17s to Trapani, Sicily on May 6th Lieutenants Burke and Bryson both of the 27th were lost. On May 10th strafing an airfield in Sicily Lieutenant George Burger was missing and on May 14th on a dive-bomb/strafing mission in the Cagliari area of Southern Sardinia Lieutenant Augustus Reese, a 94th pilot, was caught in a bomb blast and killed.[1] This occurred while dive-bombing tactics were still being developed.

As the weather warmed so did the insect problem. In May the group was hit with dysentery that hampered operations severely for a few days. The use of mosquito netting, spraying water pools, mosquito repellant and just plain common sense helped a lot. Each squadron constructed showers which, while rudimentary, proved to be exceptionally popular and good morale builders. At Biskra and at least early at Chateau d'un some of our people were still living in pup tents or shelter halves or a combination of packing crates and pup tents. Under such conditions a shower worked wonders.

On May 22nd Donald Lowe of the 94th was hit by anti-aircraft fire on a mission to Boca di Falco airfield near Palermo, Sicily. His plane was badly damaged and he was forced to bail out over the water, but was lost; and Robert Britz of the 27th Squadron was missing in action near Sardinia.

Early in June Captain John Eiland was relieved as C.O. of the 71st squadron and Captain Joe Miles assumed the post. Shortly after this the group provided escort for a York transport en route to Tripoli, later discovering the identity of the passenger, the King of England.

June 20, 1943 was a costly day for the 1st. Escorting B-26 Maurauders to Castelvetrano in Sicily, the 94th and 27th ran into heavy resistance and Lieutenants James Cronin and Edward Esser, both of the 94th were missing, later to be prisoners of war. Lieutenant Leon Thompson of the 27th, went down and Lieutenant Herbert McQown, also of the 27th, was wounded. The group accounted for a total of eight 109s in two separate engagements that day. Harold Lentz flying as 94th Blue flight leader, had one of the most successful shooting days of the war in this engagement, claiming three Me. 109s down in 45 seconds. Lentz was also shot down in an unusual setting-although not quite as unusual as his namesake 'Bugs' Lentz (they were not related).

On June 24th, dive-bombing in Sardinia, John Hay of the 94th and Lieutenants David Conn and Thurman Nielson of the 27th went down. Hay was killed in action, Nielson became a POW, and David Conn was reported missing and never returned. On this same mission Lieutenant Gilbert Butler, a 27th pilot, had the fight of his life. Coming off the dive-bomb run he was separated from his flight and suddenly found himself alone with four Me.109s diving at him. With his left engine shot out Butler headed for the deck and home as fast as he could trailed by all four 109s. As a last resort he turned into all four, aiming specifically at one Messerschmitt slightly ahead of the others. One P-38 with an engine shot out turning into four enemy aircraft was not an approved maneuver in anybody's book. By beautiful shooting, accident, luck, or all three he hit the lead 109 and splashed it. Once more he headed for Africa with his one engine firewalled now followed by three Germans. As the Germans fired, Butler jammed rudder and skidded. Once again, as a last resort, he turned into all three and downed a second 109, once again somewhat separated from the other two. This time, as he resumed his southward heading he was alone. Whether the remaining Germans were getting low on gas or just decided to stay away from this decidedly unconventional P-38 driver remains a mystery. Suffice to say Butler made it back on his remaining engine.

June 25th was another costly day over Sicily. Dive-bombing at Bocca di Falco air field five were missing including Captain Stentz, group Executive Officer, and Lieutenants Bennett, Hall and Freng, all of the 27th, and Max Rayburn of the 94th. Max was seriously injured in the right arm when a cannon shell exploded close by his cockpit and fragments hit his arm. He managed to get back as far as Cape Bon where he landed and ended up in a hospital in Tunis. He told Dr. Garretson, who came up to see him, that he hadn't been able to reach the morphine surrette in the kit strapped to his right leg. The reply was, "Just as well; if you had you probably wouldn't have made it to Cape Bon."

Late in June the 1st Fighter Group moved once again, this time to Mateur #3. The new location was an airfield on a dry lake bed south of the town of Ferryville in Tunisia and about a mile north of the city of Mateur. A railroad spur runs through the camp dividing the main camp area from the line. The Mateur location was about as close to Sicily as one could get without swimming. Sicily was some 100 miles to the northeast. With the invasion date coming closer the site made sense. It also made for more frequent missions to Sicily up to and including two a day during early to mid-July. While the airfield at Mateur was on a dry lake bed it was by no means desert. Much of the surrounding area was or had been farming country. In the distance to the northwest was a hill mass named Garaet Achkel. A large prisoner of war camp was located at the base of the hill. Mateur was very hot on the dry lake bed during the summer months. The lake bed was split by large cracks up to two inches in width and very uneven which made walking difficult. Insects were particularly thick in the area and the civilian population was almost 100% infected with malaria which meant atabrine or quinine for all hands. Atabrine gave everyone a peculiar yellow-green skin color.

On the 1st of July Major Otto Wellensiek took over as the new commanding officer of the 94th. Major George Sutcliffe, an original, left for the States nearly a year from the date the group had departed the USA.

1. Reese was from the town of Shallowater, Texas; Reese AFB in Lubbock was named in his honor.

L - Lt. Charlie McCann. R - Sgt. Gene Pillard, 71st sqd., Mateur, July 1943. (McCann)

Lt. William Royer, 71st Pilot, Mateur, July 1943. (Hansen)

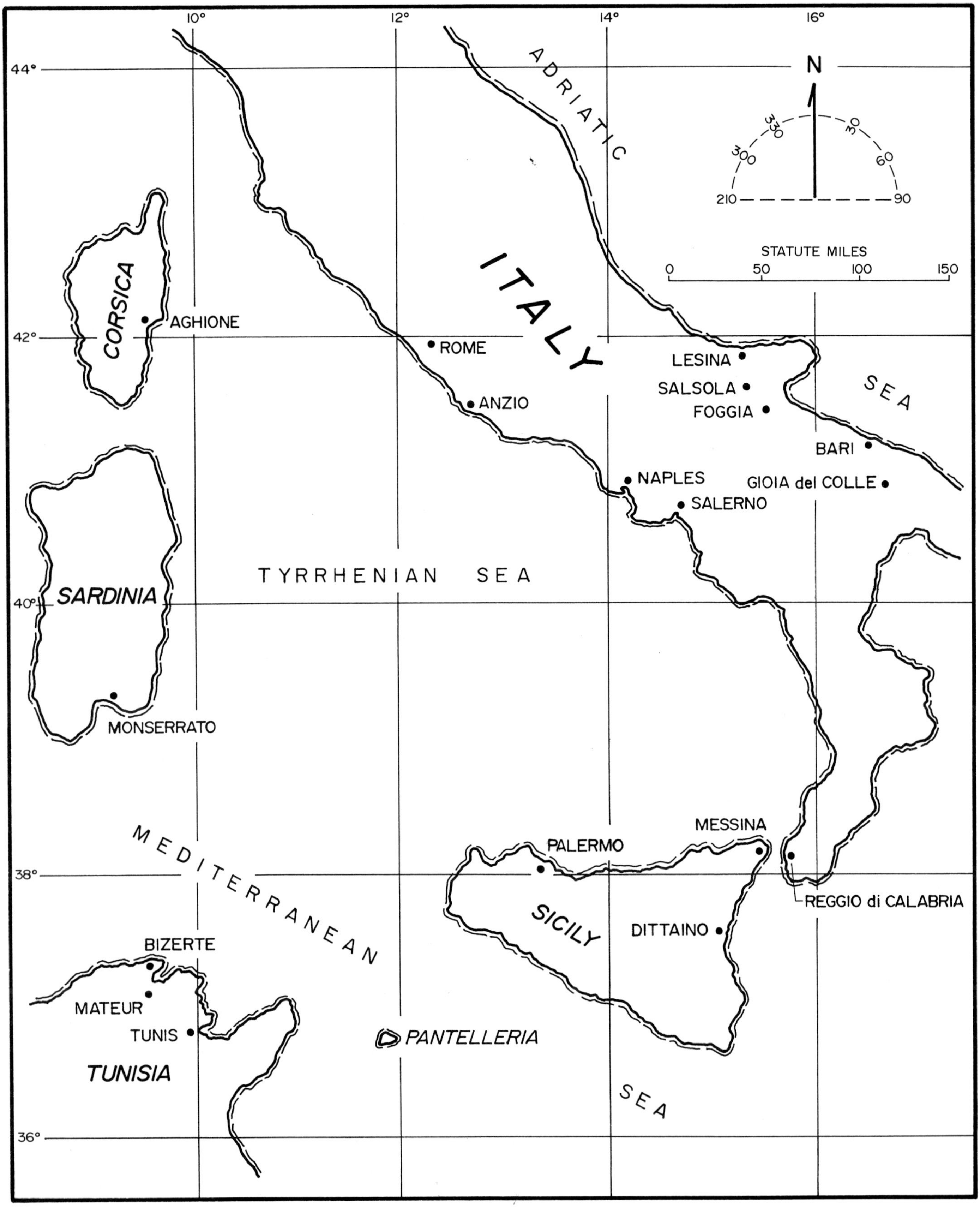
10°
12°
14°
16°
44°
42°
40°
38°
36°
ADRIATIC
SEA
N
330
30
300
60
210
90
STATUTE MILES
0
50
100
150
ITALY
CORSICA
AGHIONE
ROME
ANZIO
LESINA
SALSOLA
FOGGIA
BARI
NAPLES
GIOIA del COLLE
SALERNO
TYRRHENIAN SEA
SARDINIA
MONSERRATO
MEDITERRANEAN
SEA
MESSINA
PALERMO
REGGIO di CALABRIA
SICILY
DITTAINO
BIZERTE
MATEUR
TUNIS
TUNISIA
PANTELLERIA

8

SICILY AND THE ITALIAN BOOT

Following defeat in Tunisia the Luftwaffe underwent a considerable reorganization. Luftflotte 2 controlled Italy, Sicily, Sardinia and Corsica. Fighter strength was increased from 190 in mid-May to 450 in early July. Close to 40% of all German fighter production from May through July 15th went to the Mediterranean. Senior staff also saw changes with Wolfran von Richtofen taking over as Luftflotte 2 commander and Adolf Galland controlling fighter operations in Sicily. The impact of these changes would be seen in fierce fighting for the 1st Fighter Group and indeed for all Allied air operations beginning with Operation HUSKY (invasion of Sicily) and mounting through the month of August 1943. A point in our favor in all of this was Hitler's irrationality with regard to withdrawals. By forbidding any withdrawal planning the Luftwaffe squadrons were lucky to escape with their personnel. Much of their equipment and spare parts had to be abandoned both in Tunisia and in Sicily.

There was no serious Luftwaffe opposition to the landings in Sicily but nearly everything else went wrong particularly with the airborne operation. Faced with high winds, flak (ours and theirs) plus smoke from earlier bombings, only 12 gliders landed in the designated area. Sixty-five were swept into the sea drowning the heavily laden troops of the British 1st Airborne Division. The remaining 56 gliders came down all over southern Sicily. The followup drop by the U.S. 82nd Airborne was equally scattered. Fortunately the sea invasion forces had better luck.

Operation HUSKY began on July 10, 1943. 1st Fighter Group flew eight separate strafing and dive bombing missions to eastern, southeastern and central Sicily in early support. From twelve to eighteen P-38s were in each mission as they dive-bombed and strafed railroad and highway bridges, enemy vehicles, roads and buildings. Two of our pilots were missing on the 10th. Lieutenant Robert Kuba, a 27th pilot seen to crash into an enemy tank which he was strafing and Howard Gilliam, a 94th pilot, was also killed.

Lieutenant Robert Viall, a Cragmore pilot, kept a little loose leaf notebook diary, and describes a Sicily strafing mission:

> "Went out on the deck. Climbed to 6,500 feet after sighting coast, dropped belly tanks just before crossing coast. We took our time and picked out a nice bunch of trucks. We bombed the — out of them, then peeled off in flights of four and went down and strafed the column of trucks. One right in front of me started to pull off into an orchard. I really filled that poor truck full of 50 cal. and 20 mm shells. I got one more truck and then strafed a railroad station...I got some machine gun bullets in my left intercooler and a hole in one prop. Got back with no harm except to the plane."

On yet another dive-bombing mission, on July 11th, four pilots were missing. Lieutenants Chapman and Diamond collided in mid-air. Chapman later returned after working his way through German lines to reach Allied territory. Robert Viall describes in his diary what he called the resurrection of Lieutenant Chapman:

> "If no one had told me before hand I would have believed I was seeing a ghost walk into our tent. It was Chapman all right but he was very thin and worn from his fatiguing experience.
>
> "It all happened so quickly (Chapman and Diamond's mid-air collision) that no one knew exactly what had occurred. Chapman's flight was being attacked by enemy fighters. A turn to the right was called, but somehow a delay was encountered when the flight leader feinted a left turn and then swung into a steep right turn. It was in that turn the Chapman and Diamond collided. Some of the fellows who observed the accident stated that Diamond was on fire in a vertical dive. Chapman was last seen in a flat spin.
>
> "Looking back out of his plane after the crash, which Chapman thought was due to enemy fighters on his tail, he was surprised to see that his tail boom was missing. Immediately he took to getting out of the cockpit. After what seemed like ages, while he pawed at his safety belt and canopy, he rolled off the wing and yanked on the rip cord. The parachute opened and none too soon, landing him in a fig orchard.
>
> "As he touched the ground he could hear a car door slam and voices nearby. In his haste to get away he left behind some necessary equipment from his escape kit. He ran away quickly and came upon a shallow gully filled with bamboo shoots. He dove into this concealment and lay for what seemed like ages not daring to breathe. The thought that soldiers would be canvassing the area and might stick bayonets into his hiding place made him very uneasy.
>
> "After resting for some hours he reached for his canteen only to find that it had broken loose when his chute opened. He had neither food nor water

Robin Hansen, 71st pilot in cockpit of wrecked German Ju. 52 transport aircraft, Tunis, August 1943. (Hansen)

Tunis airport, August 1943. A wrecked Ju. 290, a rare four-engined German transport. (Hansen)

but he did have his escape purse which contained a map, a compass and some money. The compass proved to be the most valuable, and when night fell he began to guide himself back to our lines.

"Travelling by night was hard and tiring. The terrain was mountainous and very rough in spots. While crossing the valleys he ran across unripened fruits and vegetables. He did manage to eat some berries and drink from stagnant pools of water. At one point he was confronted by a farmer who found him hiding. The farmer thought him either German or English. He didn't believe that the United States was fighting against them. By hiding by day and travelling by night he managed to get close to our lines until finally on the third night he spied a Limey truck convoy. He hailed them and jumped on the running board of a truck only to find six guns aimed at his face. The rest was just a matter of identifying himself which he did with his dogtags. Chapman returned to the 71st squadron on July 16th."

On the second pass of this late afternoon mission Dee Johnson of the 94th was hit by antiaircraft fire from tanks. Although wounded he started for home escorted by Albert Fell. Well out over the Mediterranean he began to wobble, gradually pulled his ship up into a stall, then plunged into the sea.

Charlie Brown, a 27th pilot, was also hit by enemy fire and he bailed out over the target area and had a bizzare encounter with Italian soldiers. His story is as follows:

"After I bailed out (on July 11th) and hit the ground I threw my chute and Mae West into the flames of my burning plane which had crashed nearby. I heard bullets whizzing by my head and realized I was being pursued so I sought the shelter of a nearby woods where I spent the night. In the morning I went out on the road and passed an Italian sentry by giving him the Nazi salute and a Heil Hitler. Shortly after that I was picked up by a group of Italian soldiers who walked me ten miles to an Italian Army post where, after being questioned a great many times, I got partial treatment for my badly burned arm and face. I remained there all night and then in the morning I was taken in to Catania by truck.

"Upon arrival there I was turned over to the Italian Air Forces HQ. There I was given food and water. We had an air raid shortly after that and I was told that the HQ was going to evacuate to Italy. I was then taken to Messina. Knowing that if they got me over to Italy my chances of escape would be very slim, I complained of my arm and asked them to take me to a hospital, which they did. I was put in the Naval hospital there where I met with an Italian Baroness who was an International Red Cross worker. She treated me very kindly and I had good care though the food was poor. Stayed in the hospital for two weeks during which time we were frequently bombed as the hospital was situated right near the beach where a lot of the landing barges going across the Straits landed and disembarked.

"When the Allies started bombarding the city with artillery fire we were evacuated to a haven up in the hills overlooking the Straits. There I met three British officers who had been captured at Catania. Together with them I persuaded the commandant of the hospital to surrender to us, telling him that he wouldn't be in such a difficult position when the Allies finally captured the place. He consented to this and ordered the personnel to turn over their weapons to us. We, however, already had all their guns which we took while they sought shelter in an air raid. Finally, on August 17th, an American Medical officer and his staff arrived and took over. The next day we were on our way back to Africa on a C-47 and on arrival I was put in a hospital just outside Mateur. It was from there that I finally contacted my friends at the 27th."

On July 16th Otto Wellensiek, the 94th C.O., and Archie Jackson were ordered to deliver what turned out to be a set of invasion maps to Seventh Army headquarters in Sicily. The two of them flew over, found a bulldozed landing strip on the beach at Licata in southwestern Sicily. They dragged the beach, decided it looked OK and landed. The Army people who met them took them several miles inland to General Mark Clark's headquarters. On their return trip to the beach they were stopped and treated royally by the Sicilian people who gave them melons, lime sherbert and almonds, according to Archie Jackson who described it as a feast. The Sicilians had been roughly treated by the Germans and really welcomed the Americans.

Beginning in mid-July Allied air attacks began more and more to include Italy in addition to missions to Sicily. On July 17th there was a massive bombing raid on the Naples area. The docks, railroad marshalling yards and industrial areas were hit hard in order to interdict the supply lines to Sicily. Given Italy's mountainous terrain many of the German supply shipments had to funnel through or near Naples.

About this time the 71st Squadron history recorder noted an unusual food development. As he put it, "Some quartermaster officer in the rear echelon slipped up very badly. We got in some fresh meat - stew meat - and even got hold of some ice. The campaign is practically a bed of roses with a deal like that. Think of it, fresh meat."

On July 19th Rome was hit for the first time. The raid was planned with particular care and, as in Naples, concentrated on the outlying rail yards and Ciampino airport. Leaflets were dropped ahead of time warning the civilian population and non-targets, i.e., Vatican City and other churches and antiquities.

The decision to bomb Rome was based of necessity on political as well as military considerations. Militarily Rome was the key to the Italian system of communications and transportation excepting that on the far east coast to Foggia and Bari. Politically the Rome attack was expected to widen the gulf between Mussolini and the Italian people. Nonetheless Rome was an extraordinary target. The bombing was to focus on the two marshalling yards in the suburbs, the Littorio and the San Lorenzo and Ciampino airport; nothing else would be targeted. The overall mission was led by the 97th Bomb Group with General Doolittle flying co-pilot in one of the aircraft. Our mission report notes that the bombing pattern appeared to be good with fires started in the dispersal area and near the stores building. Rather surprisingly it goes on to say that numerous other small fires were observed around railroad yards and in the center of Rome! In fact there were a few stray bomb strikes but no serious damage to civilian centers.

On the 22nd the same treatment was meted out to the rail yards at Foggia on the opposite side of the peninsula. On the Rome raid we escorted B-26s although both B-17s and B-24s took part.

Lieutenant Robert 'Smokey' Vrilakas, one of the 94th replacement pilots, provides a good description of what passed for 'relations' with some of the rear echelon types in what was supposed to have been the same Army. In Africa it was the Atlantic Base Section. Later, in Italy, it was the Peninsula Base Section-probably the same people.

"During our stay there (this was in transit to join the 1st) we roamed back into Casablanca a few more times, there being not much else to do. Since we were transient officers we found ourselves to be pretty much ostracized from the permanent party

Low pass prior to landing, squadron unknown, Mateur, 1943. (McCann)

military around Casablanca. We were barred from all the support unit clubs and got quite a bit of harassment for not saluting every First Lieutenant and above we met on the street. They just couldn't bring themselves to accept those "unmilitary" looking officers, wearing Brazilian boots, leather jackets and crushed hats—the distinguishing marks of a fighter pilot. Considering the quality of food and living conditions at Berchid we strongly suspected that anything other than C-rations and Spam was being siphoned off by the support personnel who lived in requisitioned hotels in Casablanca. We never found out because we were never allowed in their clubs. In a short time those members of the Atlantic Base Section had made themselves just about as popular with us as the enemy. For kicks we would sit at a sidewalk bar and sing, 'Mother take down your service flag, your son's in the ABS'. This was not considered a positive contribution toward good relations with members of that organization, and we were all but barred from the city."

On July 15, 1943 the 94th was ordered to dive-bomb a bridge northwest of Catania not far from the slopes of Mt. Etna. This was a 94th mission only; the briefing was short and time was urgent. Jack Hanton was flying Blue flight Tail-end Charlie, Lieutenant Holcomb was element leader and Lieutenant Boggess was flying Blue 2. Jack Hanton, behind everyone, noted that Holcomb was doing some erratic flying in and out, up and down, which was certainly unusual for him. On arrival at the target the flights slipped into echelon right. Red then White flight rolled into their dive-bombing runs, and then it was Blue flight's turn. Just as they were beginning their dives Holcomb slid over and into Boggess cutting off both tail booms on the #2 plane. Both dove straight into the ground. Jack Hanton dropped his bomb, pulled around and saw nothing but two smoldering hulks of what had been two comrades and two P-38s. What happened could neither be explained nor understood, certainly not by the other pilots who simply couldn't believe it. The flight surgeon did however and, in retrospect, pilot fatigue or combat fatigue apparently was the culprit. Holcomb was certainly an experienced pilot, in fact it was unusual for him to be flying in the number three position.

The July 15th mission was the first for 'Smokey' Vrilakas. His recollection of settling in to the squadron, the first briefing and the first mission are well done and in many ways typical of most of us on our first mission. 'Smokey' concentrated on doing his job as Red #2 to the extent that he was unaware of the loss of Boggess and Holcomb until the 94th returned to base. He was chagrined at this but it is certainly not surprising. On one's first mission, nearly always as Red #2, you stick to the Red leader like glue and do exactly as he does or says to do. Smokey's description follows:

"Summer had arrived and it was hot and dry with more than a little dust blowing about. Although one gets used to it the complete lack of color in an overseas combat outfit was at first depressing. Everything including our clothing was light tan or olive drab. Our aircraft were camouflage painted, of course, and any support equipment to include vehicles was likewise painted or under camouflage netting.

"The first few days were spent in getting settled in and drawing our combat gear. Besides flight clothing and helmets we were issued a parachute, Mae West, .45 caliber pistol, oxygen mask, dinghy, canteen and a survival kit. (A Mae West was an inflatable vest to keep one afloat). The survival kit, along with the dinghy, was attached to the parachute pack. The dinghy made up part of the parachute seat and had cans of fresh water, flare, etc enclosed. All were necessary items but when you sat on them for hours you could feel them right through the protective cushion.

"Pilots scheduled for a mission were notified the night before and given the time of mission briefing-usually early in the morning. Either the squadron intelligence officer or the flight surgeon would awaken you in time for breakfast etc. I don't think I slept too well that night and breakfast of powdered eggs wasn't very appetizing. Briefing consisted of the designated target (usually shown by string on a large wall map), enroute time, routes, altitudes, alternate target, anticipated enemy ground and air opposition plus expected weather enroute, at the target, on the return and at home base. Other information included rendezvous or meeting time with the bombers, if applicable, radio frequencies, call signs, form-up procedures and armament to be carried. On occasion a special briefing by the group commander might follow. Because of possible compromise of classified information we couldn't write much on paper. We often wrote essential information like start times, take off and target times on the back of one hand where it could be easily smudged or rubbed off if need be. Finally we were given a time hack and issued escape packets. These were sealed and carried in a flight suit pocket. They contained foreign money for the countries we would be traversing plus a small U.S. flag and a map of the area. Not to be left out, the group chaplain would ask a blessing and we would be on our way by Jeep or weapons carrier. We would be driven to personnel equipment to pick up our parachutes and oxygen masks, a latrine stop and on to our respective aircraft.

"It was a general rule that a new pilot flew his first few missions as Red #2, wingman to the assigned squadron flight commander (Red #1). That was the most protected position and the easiest to fly. From there he would be assigned to element wingman also known as "tail-end charlie"—the most vulnerable position because you were the farthest back in the formation. The squadron operations officer, Major Hagenback, told me that day to just fly good formation on the leader and do everything he did including dropping my bombs when he dropped his. We had to maintain strict radio silence even for take-off so as not to compromise the mission. Generally there would be little or no radio transmissions between aircraft until we sighted enemy aircraft or had reached the target area. On the return from the target it was a different story. The enemy knew we were there and had us on radar so there was nothing to compromise.

"Each aircraft had an assigned crew chief and each pilot was assigned his own aircraft and crew chief. My crew chief was a young Staff Sergeant from Ohio named Chet Bala. Two 500 pound bombs (one under each wing) were set up for my first mission. I had never dropped a live bomb before. When attached they add a small element of risk in that they could be inadvertently dropped during taxiing or take-off. They were designed so that they would not be armed until dropped a certain distance however. My first experience in watching an armament crew kick a couple of bombs off the back of their truck on to some rubber tires gave me a bit of a shock but it was routinely done without serious incident even though not a recommended practice.

"After preflighting the aircraft it was time to get set in the cockpit and be ready to start engines exactly on the time briefed. Chet Bala would always climb up on the wing to help me get settled. The Mae West, unlike later styles, was bulky and made of stiff rubberized canvas that managed to rub your neck raw before the completion of a mission. Between it, the parachute with dinghy and survival kit attached, the .45 pistol and water canteen, it was awkward getting in and strapped down in the cockpit. Something was always catching on part of the aircraft. There's always a feeling of mounting nervous tension at that point and it seemed easy to fumble at every latch or snap that had to be attended to. It's particularly worse on your first mission but it is always there. Fortunately, once off the ground things improve immensely. But anyone who says there isn't a scare factor present is kidding himself as well as the person he is trying to convince.

"Chet crouched on the wing and talked to me about the airplane and I could tell he knew it intimately. It helped to fill the time of growing tension and he stayed there until both engines were started and running smoothly. He did that on every mission and would slide off the wing, latch the ladder[1] and go out in front waiting for my signal to marshall me out of the revetment to the taxiway. He would always

Dick Lee (L), Bob 'Smokey' Vrilakas (R), 94th Sq. (Lee)

be there to meet me on my return - 'sweating me out' it was called.

"The squadron leader started taxiing out so I moved out to follow him to the end of the runway. At that point we both went through our pre-take-off engine checks and procedures; then on a green light from the tower, lined up on the runway in much the same position as we would be in flight. The runway was too narrow to permit take-off in pairs so as soon as he applied power I did also, holding the aircraft with brakes until he started to rotate for lift-off. Once in the air he would start a slow turn which I would follow in a sharper turn inside to catch up. He would continue the turn until the rest of his flight joined up and eventually the other two flights. This same turning pattern would continue, if other squadrons were involved, until the entire group was assembled.

"I closed in tightly on Red #1 and immediately after the squadron formup was completed we turned north climbing enroute to about 8000 feet. We passed over Tunis and then continued out to sea over the Mediterranean. The island of Pantelleria appeared, but I was too busy keeping tight formation to enjoy the scenery. After about an hour and a half flight time there was a glimpse of land in the distance and for the first time I saw enemy territory. Radio silence prevailed as the squadron leader began a descent toward the shore and target. His job was much more complicated as he had to navigate very accurately and position the squadron so that we could dive-bomb without having made our position known to the enemy any sooner than necessary. At the last moment he broke radio silence, telling us to arm our bombs (an arming wire controlled from the cockpit readied the bomb for release) and fly echelon right. An echelon formation made it possible to follow into the target [a bridge at Catania, Sicily] on pretty much the same path.

"There was obviously a heavy battle in progress, as we could see numerous fires and smoke, and haze in the target area was heavy. I heard my flight leader announce the start of his run and followed slightly in trail as we heeled over and started a steep dive. At his command to 'Drop bombs - now,' I pressed the bomb release button. They were on the way and I had fired my first shot. The radio was then filled with indistinguishable chatter between aircraft as we started to climb back up, reform and head for home. There were a few black puffs of flak appearing, but surprise had been on our side so that we were in and out before the ground fire could get very well organized. (The German 88mm cannon was an excellent weapon. Unlike our 90mm the 88mm could fire horizontally at ground targets and/or vertically at aircraft. When concentrated for air defense it could be devastating, particularly to bombers since they couldn't maneuver enough to provide effective countermeasures. Fighters could lessen the threat considerably by rapidly changing course and altitude).

"We proceeded directly back to base, relaxing by flying a loose formation then tightening up on approach to the base for landing. It was a morale boost to the ground troops and a matter of pride to the pilots to look as sharp as possible at that point. We always approached the runway echeloned to the right at an altitude of about 200 feet. Over the runway we'd do our peel-up and land out of a tight circle. Sergeant Bala was there to marshal me in to the parking spot and was on the wing as soon as the engines were shut down helping me get unstrapped and out of the cockpit.

"We were picked up in Jeeps and taken directly to the debriefing tent where the group and squadron intelligence officers wanted to know everything about the mission. It wasn't till then that I learned that two of our pilots (Boggess and Holcomb) in another flight had collided over the target and gone straight in with their bombs. I was a bit chagrined that it had escaped my notice but the squadron commander assured me that I had done exactly as he had wanted me to do — stay right in formation and concentrate only on that. So much for the first mission, but I told myself from there on I was going to take in more of the big picture. The flight had lasted three hours."

Robert Viall included a close call in his diary for July 22nd on what began as a relatively easy trip escorting a PBY on an air-sea rescue mission. The destination was far up the northeast coast of Sardinia. After about four hours flying time Viall switched away from his belly tanks only to discover that a lot of gas had siphoned out of his internal tanks. His left reserve tank was empty and about 25 gallons were gone from his left main tank. He started for home following the coast of Sardinia flying on the deck. He finally sighted land with about 30 gallons of gas remaining. Climbing to get over the barrage balloons around Bizerte, both engines coughed and quit. Luckily there was an airport in sight so he dropped his gear and came in for a deadstick landing after seven hours in the air. The PBY rescue mission was successful as well.

July 1943 closed on an oddly positive note with the saga of Lieutenant John (Jack) Hanton and his single handed escort of a B-17 bomb group. Jack was definitely a "new guy" as he termed it. He joined the 94th squadron at Chateau d'un with the grand total of 12 hours in the P-38. He had never fired the guns and never feathered an engine. Nonetheless, on one of his early missions he escorted the 99th Bomb Group to Naples, shot down an Me. 110 and got back, all of this by himself. On this particular mission Jack was the only one in his tent scheduled that day and for whatever reason he wasn't awakened in time for the pre-mission briefing. He tells us what happened from there on:

"It was only when the squadron pilots had returned from the group briefing and were getting into and out of Jeeps, etc. that their voices awakened me. I jumped up, put on my flight suit, canteen and .45 and ran out, got my chute, jumped into a Jeep and asked one of the pilots where we were going. By then we were at our planes and he said, 'Just follow along and you'll get there'. There was a snafu of some sort after 'start engines', and we shut down for a few moments adjacent to the runway. When we started up again I couldn't get one of my engines going (there's this new guy again) - must have flooded it. Anyway my good old crew chief came over in a Jeep and we got her going again. By then the whole 1st Fighter Group was off and over the hills, nowhere in sight. I figured a P-38 group of aircraft would be easy to spot and I remembered we usually picked up our bombers off the shoreline of Bizerte or Tunis. I saw some type of aircraft in the distance and caught up to them when they passed the shoreline between Bizerte and Tunis. As I flew closer I could see they were B-25s, which we never escorted, so I kissed off and began circling and climbing upward on a northeasterly heading. In the distance I could dimly make out a large number of aircraft and lit out for them. In 15-20 minutes I caught up with a group of B-17s. But where was their P-38 escort? Maybe the P-38 Leader (another group naturally) missed the bomber rendezvous point and would be along at any moment.

"I stayed with the B-17s as they continued their en route climb to higher altitudes. This 'new guy' was at least smart enough not to encroach into the B-17s range or angle of machine gun fire. This necessitated wide S-ing from one side of the bomber stream to the other and from the lead bomber on back to the last aircraft. I always wondered what those B-17 crews thought about that lone P-38 out there escorting them? Never! Must be a German pilot with a captured P-38.

"As the B-17s made landfall over Naples and still climbing - I believe we were around 18,000 feet then - and I had banked over to the right side of the bomber formation, I looked down and started counting the large concentration of ships and flying boats in Naples harbor. I thought how great it would be to report all this valuable information to our intelligence guys when I returned. Whatever it was, fighter pilot instinct or the Good Lord looking out after 'dumb new guys', I managed a fast check of my 7 o'clock position, and what I saw made me turn white with fright. For diving out of the sun was the biggest, meanest looking Me. 110 I'd ever seen! In fact it's the only one I'd ever seen! And, even afterwards, never saw another one. The big, black 'Iron Cross' on the side looked huge, and, I remember as I dropped my nose, hit combat flaps, advanced rpm and power, turned on the gunsight and racked my P-38 around in a screaming hard right turn, I could see those four cannon in the nose, plus whatever, and all that I had were four .50's and a 20 mm. Lord, help me now!

"I still had my damn drop tanks on, and was hitting a high speed stall as I buffeted around in my 360 degree turn. At least as long as I kept the Me. 110 out there, he couldn't hit me. And, as I continued to rack her around I actually got in a good position to fire a long burst across his left engine and wing. The Messerschmitt appeared to stagger and to fall off. I proceeded to continue my hard right turn, descending fast, finally dropped my tanks and pressed on out of that area but fast. If those B-17s needed fighter escort someone else could provide it - not me!

"Once clear of the area and half-way home I called 'Big Fence' for a steer and got back in one piece to Mateur. When I returned to the 94th my ground crew noticed the white tapes were off my guns and, after engine shut down, that my right engine propeller hub had a nice clean bullet hole through it, which really now was the only evidence I'd encountered anything, let alone a group of B-17s and an Me. 110. Here was this 'new guy' (up to that time I'd never even fired the guns on a P-38) going out by himself, escorting a group of B-17s and claims he's jumped by an Me. 110! Oh well, at least the Intelligence troops said claim a probable or damaged.

"Retribution came a day or so later, on what I believe were 'INTOPS' (Intelligence Ops Summaries); the 99th Bomb Group reported that the Me. 110 claimed as 'damaged/probable' over Naples/Capri is confirmed as destroyed. A further report through friends some days later said, 'Tell that P-38 driver over Naples the other day when he shot down that Me. 110, it's good he got out of there fast, there were seven more of them coming down after him"

Frank Maglio, 71st pilot, Tunis. (Hansen)

Jack completed his missions and returned to be an instructor at Van Nuys. One of his students would be Dick Sykes a 1944 94th pilot. Not only that, Jack's P-38, somewhat the worse for wear, also turned up at Van Nuys. Incidentally the rest of the group, which Jack couldn't find, escorted B-26s to Salerno. All returned safely.

Smokey Vrilakas recalls one of the more unusual calls to duty that took place in late July. It may have been interesting for the pilots but it must have caused a certain amount of apprehension among the crew chiefs. It seems that 1st Group Headquarters passed the word that the maintenance people were having a difficult time of it because of low manpower. All pilots were directed to go down to the flight line during non flying days and help their crew chiefs in any way possible. Sergeant Bala looked at Lieutenant Vrilakas somewhat dubiously when he "reported for duty" and after some thought (and probably no little concern for the airplane) said, "Well you could clean the landing gear oleo struts."

"So with a bucket of solvent and a rag I went to work," noted Smokey. "The crew chiefs no doubt got a kick out of putting all those officers to work but that kind of teamwork was good for morale and helped get the job done. We did that for a week or two then fell back into the old routine."

Continuing the Sicilian campaign, on August 5th Lieutenant Harold Herr, the victim of a malfunctioning compass, went far off course and had to bail out off the coast of Sardinia. He spent some three weeks in a dinghy, was finally picked up by a passing freighter and brought to North Africa; weight down to under 100 pounds. Herr returned to the United States and became a very believable instructor in a survival school.

Tunis, the capitol of Tunisia, close to Mateur, lay in ruins after Axis forces had been defeated. August 1943. (Hansen)

On August 9th two missions involving all three squadrons escorted B-26s to Angitola. On withdrawal three enemy fighters attacked the 71st but were driven off. Not so with the 27th. They were attacked and two ships were damaged so badly that Lieutenants Peterson and Graham had to make emergency landings in Sicily. Both men returned the following day leaving their ships in Sicily.

August 10th proved to be yet another costly day this time for the 94th. This was another dive-bombing/strafing mission near Bova, Italy carried out by the 71st and 94th squadrons On this mission we lost Lieutenants William Grieshaber, Glenn Terry and John Wilson all 94th pilots. Terry was hit, crashed and exploded on the railroad tracks, Grieshaber was hit, caught fire and crashed and Wilson went down in the sea. An unconfirmed story has it that Grieshaber was captured by the Italians, turned over to the Germans who shot and killed him. Both events, while certainly possible, were unusual. The Italians, either troops or populace, seldom willingly turned Americans over to the Germans at this stage of the war. Ordinary Germans troops, in turn, rarely shot captured Americans out of hand at least at that time. In any case Grieshaber did not return and was not a prisoner.

Two days later, after attacking Grazzianise Airdrome in Italy, Lieutenants Miller and Peters of the 71st Squadron were forced to bail out over the sea after tangling with some twenty-five enemy aircraft, apparently in two separate actions. Lieutenant Viall described the first part of this fight:

"The 71st was the lead squadron and were with the bombers bringing them out of the target area and out to sea. About 20 miles out to sea the 27th called to us to come back for they had been jumped. We turned back into the fighters while the 27th continued on with the bombers.

"Then the fireworks really began. There were Me.109s and FW 190s all over the place. They came in from so many directions that our squadron got split up. I was leading the second element in Blue flight when we were attacked by five 109s. They were called in before they were close enough to shoot and Lieutenant Berens called for a break. He called three times and the third time he screamed, 'Jesus Christ, break right!' All this time I could see the 109's sitting on our tail winking at us. I wanted to break away by myself for I thought surely they would get our whole flight. I was scared to death. Finally the turn was made but it was too late. Lieutenant Peters in front of me was hit and rolled down on top of me. I had to kick out from under him and thus was left out of the formation by myself. I immediately reefed her into a tight turn to the right for I saw another one of our flights breaking that way. A lot of company is desirable in such a situation. I was going to tag onto this flight when I saw four 109s coming down on their Tail-end Charlie.

"The second 109 was very close to me when he saw me and broke head-on into me. I squeezed the buttons in a very short squirt. The 109 blew up and pieces flew everywhere. At first I thought my canopy was on fire but it was the flash from the exploding Me. 109. It seemed to me we would collide head-on but instead I flew through the wreckage and a piece of the plane tore a hole in the intercooler of my right wing causing a drop in manifold pressure. I pulled down combat flaps to turn back into the flight and then saw the other 109 perched on the tail of this flight. I gave him a short burst and he rolled up on his side with black smoke pouring from his nose. The last I saw of him he was hanging in that suspended position."

This fight led to an interesting discussion of tactics by Dick Catledge:

"The squadron of twelve, three flights of four kind of relaxed and all of a sudden four Me. 109's hit us from the rear. Well we had twelve, three flights of four. So around and around we go. They weren't flying in a flight, they were flying individually. Everywhere I looked there was a Messerschmitt on a flight of four's tail. You would look around and there was another flight of four with one on their tail. Go chase him off and then you got one on your tail and someone chases him off yours. The thing is we were staying in flights of four and, in effect, we were outnumbered by these four individuals. In the process my tentmate, Roger Miller got shot down. When we got back I was really ticked off that four airplanes could give twelve so much trouble. What I couldn't understand was why we weren't permitted to break up into elements of two. Then we would have had six elements against four and had them outnumbered. But we were strictly prohibited from doing that. You are supposed to stay with your flight leader at all times regardless.

"So, when I got on the ground I went to my operations officer, Lee Wiseman, and talked with him about it. I thought that we ought to change our tactics particularly in a case like that and break up into elements of two. Well he didn't say he agreed or disagreed or he didn't have the authority to say anything so I went to the squadron commander, Captain Joe Miles, and I made my pitch to him about tactics and recited what had happened and how we were outnumbered. I didn't understand and wanted to know why we had to stay in a flight of four when it created so many problems. Well, he didn't know the answer, so I said 'I'd like to talk to the group commander, Colonel Robert Richards. Which I did.

"I made the same pitch to him. I was really upset. Primarily because I had lost a tentmate and classmate, but it just seemed so stupid to be so dumb as to sit there in a flight of four. Here I am a second 'Louie' just out of flying school complaining about the tactics we're using. He was very tolerant and explained to me the reason that they insisted on the flight integrity of four is that when newcomers came in flying with an experienced flight leader, staying with him was the best chance of survival and gaining experience. You would start flying on the flight leader's wing, then go back to number four man on the element leader's wing, then progress to number three man and finally to flight lead. That was his rational for maintaining the integrity of the flight.

"It still didn't make sense to me why tactics couldn't be developed to break your flight up and

Colonel Ralph Garman, a charter member of the 1st Fighter Group, was reassigned in August 1943. (Ilfrey)

increase your numbers thereby. We never did and I don't know that anyone did in World War II."

In fact the flight of four ships was originally considered as a flight of two, two-ship elements. Early in the war the two elements sometimes flew abreast of each other for better mutual protection. Logically the number two man was the least experienced and flew on the leader's wing. He then moved to 'tail-end Charlie' a position with a good bit of flexibility leading to the element leader position which, by the same logic, should have included flexibility and responsibility (for a wingman). In fact the flight of four became inflexible as Catledge found out and more often than not became a flight leader/shooter and three guys flying formation, granted that number four was usually weaving around in the back.

On a more positive note a goodly number of our people went over to the P-40 field on the other side of Mateur and enjoyed an excellent show put on by Bob Hope, Frances Langford and company. A couple of people in P-38s gave one and all a good buzzing to climax the show.

On August 13th the Rome area was raided for the second time. This was a massive raid once again targeting railroad marshalling yards, industrial areas and airfields. Both the 1st and 82nd groups escorted over 100 B-26s on this mission. Opposition was not particularly heavy but this was a short lived respite. The German air strength was increasing probably in anticipation of an Allied invasion of Italy. In fact the Rome raid was really more in preparation for Salerno than the campaign in Sicily since Operation HUSKY would end on August 18th. On the Rome raid Donald Kienholz shot down a Me. 109.

On August 14th all hands were called up in formation to be present for General Spaatz' visit. He was accompanied by Generals Doolittle and Webster and British Air Chief Marshall Tedder. Departing Colonel Garman and the squadron commanders were all decorated.

1. The P-38, because of its size, had a little two step ladder and handle at the rear edge of the wing directly behind the cockpit. The ladder swung forward and up into the bottom of the pilot compartment when not in use.

Author's Footnote:

In an interesting aside, representatives of the 1st Fighter Group were able to review records of the French-American Maritime Hospital located at Ferryville, Tunisia. Hospital records reveal that Lieutenant Lawrence Pace of the 27th Squadron died of wounds on December 4, 1942. Records also show that Lieutenant David Everett also of the 27th Squadron was killed in the vicinity. Both were buried in the Ferryville hospital by the Germans and their graves were well cared for. Near the graves of our pilots were the graves of Germans, Italians, French and English, all of them decorated by the Germans. Some victims of the air battles were not so fortunate – if that word is appropriate – and were buried where they fell, literally by the side of the road.

Off the road to Bizerte, July 1943. It reads: Unknown British Aviator. Born ? Died In Action 12 May 1943. (Hansen)

Mateur on a rainy day. Horace Landry looking for high ground. (Krook)

Bridge at Constantine - a perfect fly under target. (Lenhart)

9

ITALY and SALERNO

This chapter, beginning in mid-August, anticipates the Salerno invasion by nearly a month but then, so did activities in the air. With the close of Operation HUSKY, Sicily was secured and attention turned to Italy. The Germans, anticipating an invasion, or more than one, had to deal with where and when. A crossing of the Straits of Messina was a good possibility and the British obliged by moving the Eighth Army from Sicily to the toe of Italy on September 3, 1943. A second plan involved a landing further up the peninsula. This would be limited by the supply route, and the range of air cover and topography, as a great deal of the Italian landscape is up and down. Locations with good beach accessiblity, and some level space between sea and mountains, were scarce. But an area southwest of Naples fitted the bill and the name Salerno was to become famous, or infamous. Yet a third invasion force would land on the heel of Italy at Toronto on September 9th.

In the brief interval between the end of HUSKY and the Salerno invasion (Operation AVALANCHE on September 9th) the 1st Fighter Group would experience some of the fiercest fighting of the war. During that short span the group would earn two Presidential Unit Citations and also lose a lot of good pilots.

During mid-August, the German air force, mostly Ju. 88s made repeated night raids on the Bizerte harbor area in an effort to break up the Allied build-up for the anticipated invasion of Italy. The raids were just close enough to Mateur to make for a spectacular show combining our own anti-aircraft fire with the German bombers high above. The Germans obviously were trying to interfere with the use of Bizerte harbor as a staging area for the invasion of Italy. They made no attempt to hit any of the airfields which were only a few miles south of Bizerte. The enemy raiders were coming from airfields in Italy and Southern France which in turn would be hit by Allied medium and heavy bombers.

During the softening up raids into Southern Italy, Lieutenant Thee Smith was missing from a B-26 escort mission. His belly tanks wouldn't release so he peeled off and headed for home apparently the victim of an errant compass judging by the radio transmissions and was captured but escaped to return later to the 71st squadron to complete his missions. A few days later Lieutenant Charles Simpson went down some eighty miles off the Italian coast, was picked up and returned after spending about forty-eight hours in a dinghy. Aside from some weight loss he was none the worse for his experience.

On August 20th on a return to the Naples area B-26s bombed Caserta and five of our escorting pilots were lost. From the 94th Robert Allan, Philip Sager, Jr. and Roane Sias were missing. Sager downed two 109s then was forced to bail out. Allan's ship broke up in the air. Sias bailed out, was captured and escaped after being imprisoned in Naples. He returned to the squadron in Foggia in 1944. Sager was recovering in a German hospital near Naples with injuries received when his plane was shot down in flames and was left behind to American forces when the Germans evacuated the hospital. From the 27th squadron Lieutenants Adolphus Webb and Harold Post went down.

The 1st Group claimed five enemy aircraft destroyed, six probables and four damaged. Our pilots reported that the Germans attacked in pairs and threes from all directions and with an altitude advantage. Some attacked in strings of seven peeling off after each attack. The enemy attacks began as our formation reached landfall, continued into and over the target and persisted out as far as Ventotene Island. It was becoming apparent that the Germans had installed one or more elite fighter groups in the Naples area. They would make their presence felt again and again.

Beginning around August 23rd the 82nd Fighter Group joined with the 1st and began briefings and low level practice runs preparing for a Foggia area raid scheduled for August 25th. The level plain surrounding Foggia was the site of some eight enemy air bases housing an estimated 200-250 planes of all types. Wide dispersal made conventional bombing, even frag bombing relatively ineffective. Enter the P-38s.

The combined effort of both P-38 groups put 140 Lightnings in the air on a typically bright, clear Mediterranean morning. This was the ultimate maximum effort, averaging almost twenty-four planes per squadron or twice our usual twelve so that each squadron flew as two. After a very large circling maneuver to complete takeoffs and positioning and with squadrons in trail formation at minimum altitude, the mission began. Overall leader was Major George Rush, 1st Group Operations Officer.

The route led out over the sea, skirting Sicily to the west and north to landfall at Praia la Mare Italy, 500 miles across the spine of Italy and out into the Adriatic at Molfetta, all at an altitude of 1,000 feet, gradually lowering to below 100 feet as the coast of Italy drew near. Turning northwest out over the Adriatic the formation approached the Gulf of Manfredonia and their landmark, the mountainous spur of Italy. With the spur dead ahead the groups altered course to about 10 o'clock and spread out to line-abreast. The 1st headed for its assigned targets north of Foggia and south of San Severo.

Surprise was complete as the P-38s swept over the area. Bombs exploded, gasoline went up in flames as did trucks. Mechanics and ground support people were caught on wings, in shops and hangars, refueling. Enemy planes crumpled in flames or exploded. In all the Germans lost at least forty-seven aircraft destroyed and another thirteen damaged; a devastating blow to the Luftwaffe at very little cost to us. The 1st lost two planes and pilots, one at least lost to intense small arms and anti-aircraft fire. Marcel Williams of the 27th was hit by flak and forced to bail out over the target and was captured. Robert Viall of the 71st squadron had to leave the formation as he could not release his belly tanks and was not seen again. The 94th lost none although many were hit. Jack Hanton strafed a train about halfway across Italy, got shot at in return, was hit and made it back to Sicily. On recap the 1st accounted for forty-three Ju. 88s, one Ju. 52 and one single-engine fighter destroyed on the ground. One Ju. 52 was destroyed in the air and ten Ju. 88s and three single engine fighters were damaged. A railroad station was strafed and left burning, and a troop train at Potenza was strafed with a good many casualties as the soldiers poured from the train. The P-38s returned to base in pairs and flights with quite a few landing at various fields in Sicily to refuel. The 82nd Fighter Group claimed another twenty-one aircraft destroyed and forty-three damaged for a loss of three planes and pilots. [1]

Such missions did wonders for everyone's morale. It was obviously a huge success and losses were minimal. Probably not realized by most of the participants was the fact that German air defenses for the soon to come Salerno invasion were severely crippled. For this achievement (some months later) the group received its first Presidential Unit Citation. Little did anyone realize that a second citation would be earned just 5 days hence as a result of one of the most vicious air battles in group history.

For the next few days following the Foggia raid it was back to business as usual; escorting B-26s to Italy, Grazzianise airfield, to Caserta, to Naples, and, on August 29th to the Battipaglia rail yards The enemy was present and aggressive on each of these raids. In several encounters the group lost only one and accounted for at least six enemy aircraft. One oddity occurred on the August 28th mission. All three squadrons escorted B-26s to Aversa and had a brief tangle with twenty-five to thirty enemy aircraft, destroying three, but reporting Lieutenants Rudnick, Winegar, Catledge, Warren and Phillips missing. Phillips turned up the next day but not the other four - one entire flight. Catledge and Rudnick were the only two eventually listed as missing on the 28th. Until Richard Catledge returned in the spring of 1944 no one knew what had happened to that flight. In reality all four were shot down over Italy, immediately off-shore in the Bay of Naples.

Richard Catledge on his 23rd mission, relates the saga of his downing, imprisonment, double escape and evasion:

"Our job was to stay close to the bombers and try to keep the enemy fighters away. You couldn't go chasing after them. At the most we would have 24 planes escorting the bombers and you might be jumped by 100 fighters or more. But they were afraid of the nose of the P-38 and all you had to do was turn toward them. They weren't that aggressive plus the bombers had some self protection; they had gunners.

"Anyway we would usually stick with them by hook or crook on the way into the target, but on the way out we would have to turn into them (the enemy) so often that we would get stuck back in the target area somewhere before we got out. This happened on every mission I flew to Italy. You would have a fight that might last fifteen minutes to one that lasted forty-five minutes. When you are in one like that you are usually completely out of ammo and low on gas.

"One of the things the enemy would try to do is make us keep turning. When somebody comes at you you've got to take evasive action to keep them from getting on your tail. The only way to do that is to turn. It was a tactic they used to try to run us out of fuel. We didn't lose too many airplanes though. We could have been wiped out if they had been more aggressive. It was kind of hairy when the mission was going to Rome or Naples or somewhere and you knew what you were getting into. It was kind of nerve wracking. You thought 'Oh gad, not again.'

"On the 28th we got hit just before we reached the coast of Italy, maybe five minutes out over the water. There were fighters all over the sky. Our number four man (Warren) got hit. His radio was OK and he told us he was on single engine. So the three of us dropped back to stay with him. Of course, the thing to do was head for home when one of your guys was on single engine. From that moment on, when these four P-38s left the group and the bombers, every enemy fighter in the sky came after the four of us. The rest of the group went on into the target area.

"Things got pretty hot and hairy and I saw one of our boys in a parachute, at some point. Of course the flight of four got all separated. You couldn't stay together but somewhere in the turning and twisting I saw a parachute which I knew had to be a P-38 pilot.

"Then just a little later I saw a P-38 go down smoking. Then much later, out of my peripheral vision, I saw a P-38 go straight in. This all started about 15,000 feet. I just happened to see him at treetop level going straight in - just like that. So that was probably my three buddies. One parachuted, one went down smoking and one went straight in (Rudnick).

"I didn't know if they had gotten out, but here I'm still up, still going around and around all by myself. My thought was, 'OK, if I can stay airborne here, the group is going to come back out through this area, they'll see me and come down and help me.' At that time I was around 7,000 feet.

"Well, I was still up and twisting and turning when they came back out. I can see the bombers in formation and the remainder of the 1st beehiving in their formation around the bombers. And on out over the Mediterranean they went.

"By that time I was over land and I looked down and for the first time really saw Italy. It was a strange feeling. Before it was all very impersonal. We flew over Italy a lot, but I never really looked. This time I looked down and saw trees and mountains and hills and fields and I thought, 'gad, that's Italy.' I kind of figured then that I wasn't going to get out of there. Sure enough, right after that I got hit in my left engine. There was a hole in that boom that I could have crawled through. So now I'm on single engine and still going round and round, once again out over the water.

"You know, you're trying to watch everyone to keep them off your tail and turning into them but with this engine gone it wasn't going too good because you can't turn as well, and you can't turn at all well into a dead engine because the good engine tends to flip you over.

"So I'm up there really busy and some guy in this little 109 came by from somewhere on my left. I never saw him coming. I was probably at about 200 MPH, maybe 250 but he must have been goin 400 MPH. He must have come down from above. He shot at me and went by me. I was in a left turn then trying to get away from somebody and he was in a steep bank when he went by me. He was in this big, big arcing turn so he had to be going fast compared to my speed. He made this big wide turn and came on around. And I'm sitting here and I can remember thinking, 'You son of a gun you're not going to come around on me again.' He is in a wide turn and, with my slower speed I'm turning inside of him. He probably didn't even see me but I'm cutting him out and almost on his tail.

"I'm sitting there with my finger just ready to press the firing button, and all of a sudden it sounded like my guns were firing but I wasn't firing them. I took my hand off and realized it wasn't me. Somebody was behind me that close. I looked back and here is a white spinner 109 just sitting right on my tail.

"The only thing I could possibly do was pull it in to the left as tight as I possibly could. If I turned the other way I was dead. So I pulled it in real hard and, of course, stalled and flipped over on my back which is what I should have done. Problem was I looked down and it looked like I could touch the water. I hadn't realized how close I was. I knew when I saw the guy on the horizon we were low but I had no idea we were that low. So I looked down and thought, 'Boy this is it.' But thought maybe if I cut the good engine and flip it back I could get back over in time and not land upside down, which would have been too bad. I was able to do just that. I cut the good engine and flipped it back over. When I did this I lost too much airspeed so there was no chance of continuing. I held it off to land in the water and the guy behind me kept up a steady stream of tracers over my right wing. I was drifting to the left because of the drag of the left engine which is why his bullets were missing me to the right. As I slowed to stall speed I had to use ailerons to keep it level. Before I hit I had full aileron and full rudder and I was still slipping to the left. Then I hit the water.

"I got out of the cockpit under water. I don't know how deep I was. The plane didn't stay afloat, and I was probably knocked out for a time because later one side of my head was all banged up and I had a deep abrasion on my shoulder. At the time I didn't know that; I didn't feel anything. Anyway I got to the top. Of course you had all kinds of stuff hanging all over you-a big web belt with a .45 automatic on it plus a canteen and assorted survival equipment and a parachute. All I did was unbuckle the safety belt and go up parachute and all including a Mae West. The Mae West is inflated by pulling two strings on the bottom of the vest to release carbon dioxide to inflate each side of the vest separately.

"I remember feeling pretty secure about surviving because I knew I was a good swimmer and I had my Mae West. So I thought all I gotta do is inflate the Mae West but I still had all this paraphernalia on me and that nearly did me in. I'm working pretty hard to tread water. If I stopped to get to the Mae West I went under. I must not have had much breath because I really couldn't stand to go under very long. I needed the air. I finally realized that I just had to get to the vest; I did and pulled and-nothing. That CO2 cylinder was either empty or not even there.

"Well I gave up. I was too tired to get to the cylinder on the other side-too aching and just too tired. I relaxed and it was a very peaceful feeling; a very calm, peaceful feeling. Your muscles ached from fighting it and it was very stressful. To relax under water and just start floating down was very peaceful.

"Somewhere in there I saw, and I can still see this image, I was in a huge vat or room; a huge thing of water, clear water with no boundaries and I could see myself floating down. At that time I thought, 'God, no one is ever going to know what happened to me.' For some reason, even at that late stage, I had to give it another try. I don't know how, having gone through that and having given up completely and gone down, that I even had the strength to do anything, but I decided I had to fight again, and I did.

"I had a parachute with a bad shoulder strap. Whenever I was in combat the dumb right shoulder strap would fall off my shoulder. It wasn't tacked in back as it should have been and it would come undone. I kept reminding myself to get it fixed when I got back from a mission and kept forgetting about it. I'm sure that when I relaxed my arms the strap fell off so that when I began to try and surface again the parachute was out of the way and my shoulders were free. When I did surface it was easier to get a breath and easier to stay afloat. I reached down and pulled the left hand string on my Mae West and it inflated. Then I seemed perfectly normal and I swam in. I was far enough out in the Bay of Naples that I couldn't tell where land was. What I used to determine the direction of land was the black smoke from the P-38 I had seen go straight in. I swam in the direction of that column of black smoke. As far as I know the Germans didn't fire at me when I was in the water but they fired at me all the way down even when I was just about to hit the water."

Richard Catledge reached shore, was immediately siezed by Italian soldiers and incarcerated in a primitive jail. Later he was taken to an Italian Army post.

"I ended up in an Italian colonel's office filled with military people. None of them could speak English. They were trying to interrogate me, making motions and talking with their hands. I knew what they were trying to find out but I pretended that I couldn't understand anything. They finally brought in a little old man who spoke pidgin English and he was our interpreter. He said, 'What they want to know is where you come from, what kind of airplane you were flying and how many were with you.' I said, 'Tell them I can only give my name, rank and serial number.' So he translated that to the colonel who came around, clapped me on the shoulder and said, 'Ah Bravo!' That was the end of the interrogation but, at that phase, if they had pointed a knife or a gun at me I'd have probably said I came from North Africa. So that ended that and they took me to a seaplane base in the Bay of Naples. It's on a little rocky peninsula that juts out into the Bay."

Again Catledge ended up in a crude improvised cell, this time above an Italian fire station.

"In a period of two or three days there were seven of us collected in there. Five P-38 pilots including myself, a B-25 pilot and a B-25 tail gunner. Two of the pilots I knew well, they were classmates, Dave Parlett and Walter Morrison. Lou Curdes was from the 82nd and David Deisenroth, another 1st Group pilot made the fifth. We were there almost a week. At night, because of Allied air raids, we were taken out and put in their bomb shelter, a huge cavern in the rock. Everyone on the base was there including all the Germans.

"We planned an escape out of that room. They would leave the window open in the daytime and come back at night and put the lock on it and the guards were always there but the guards would all go downstairs at night and outside the building; we could hear them talking. We took the screws out of the inside hinges that held the padlock in place. We figured that the night we were going to make our escape we'd take those screws out, put the window back in, put the hinges in place, put the padlock back on and lock it. The guard coming up to lock it would probably think it never was opened that day, and that's exactly what happened."

Dick Catledge and several of his companions made a daring escape during the early morning hours, stole a boat and tried but failed to get away. They were discovered before long and were recaptured by many Italian soldiers.

"They herded us back and up to our room and they were really getting emotional. They stripped us stood us up against a wall and started pointing their rifles and bayonets at us. The room was full of them. One of our boys didn't like being pushed around. He wasn't very cooperative and one of them just slammed him over the head with a gun butt and knocked him out. I just knew they were going after us with those bayonets and there was nothing we could do. We were standing there absolutely helpless. About that time an Italian officer walked in. He calmed everybody down and made some of them, but not all, leave the room. He spoke very good English. We told him what we had done and he said, 'Well I don't blame you. I would have done the same thing.' So that ended that.

"The next day they put all seven of us in a big truck with guards all over the truck sitting between each of us, as if we were dangerous. The truck took us north of Rome to a real prison camp complete with cells - two or three of us to a cell - all surrounded by a twelve foot wall. We couldn't have been there more than two or three days when Italy capitulated. There were Americans, British, South Africans, Aussies and New Zealanders there and the guards didn't know what to do with us. They opened the cells and let us have the run of the prison. That night around midnight twenty-one of us went over the wall. We had heard that the Germans were coming to take over the prison.

"The first escape was really like a movie. It was well planned and it worked as slick as a whistle and if we had gotten that boat started we'd probably have been alright. The second escape was kind of anticlimactic. They let us out, we went over the wall and got away permanently. For the next eight and a half months I evaded behind enemy lines in Italy.

"The twenty-one who went over the wall included a New Zealand colonel, an Australian fighter pilot, several British and about eight Americans. We Americans kind of stuck together. We went up into the hills, ran into an Italian charcoal maker who put us up in a sort of wooden hut. He brought us food-wine, cheese, bread, and we stayed there about a month, then got impatient and decided to head south going two by two. The first pair to leave were two 14th Group pilots, Major Bright and Lieutenant Hyland. Deisenroth took off on his own. The remaining Americans headed south guided by a toothless Italian they called 'Snuffy'. The group gradually broke up into two's and travelled separately although they usually had a pretty good idea where the others were."

The Dick Catledge saga continued with German near misses, cold, little or no food, inadequate clothing and very real dangers while working his way toward the Allied armies in the Appenines in winter. These were his experiences but they could be equally well assigned to many of the other 1st Fighter Group pilots who worked their way south in those mountains in the winter of 1943-44. While the Italians did at least look the other way when many of these men went over the wall or otherwise escaped, the Germans now controlled Italy and the German army was a very different matter indeed as Catledge explains:

"The Germans were pretty tough on Italians if they were harboring escaped prisoners. On one occasion in a little town called Valle Pietra, Curdes and I were able to spend the night in an Italian home. When I say home you can't imagine the poverty of these hill people in their little villages. But we did have two single beds that they gave us this winter night.

"The little lady came running in just about daybreak and said, 'The Germans are here, the Germans are here!' We heard shots and looked out a window. Of course, all of these villages are built on hilltops and down at the bottom surrounding the village we could see German soldiers in their great coats standing shoulder to shoulder all the way around the village. The shooting we had heard came when some young men of the village tried to escape."

Once again Catledge and his compatriots escaped and lived the fugitive existence, encountering kindness from Italian civilians while crossing paths with other Allied evaders and Austrian deserters. By the time they ran into German soldiers, near the front, Catledge and his companions looked lilke Italian refugees, spoke passable Italian and had acquired quasi-identity papers. Their papers, beards and bedraggled looks fooled the Germans. Out of uniform, however, they were in danger of being shot if discovered.

"We had been between the lines for days dodging artillery fire and geting strafed and dive-bombed by Allied fighters, either Spitfires or P-51s. At that time we weren't sure what a P-51 looked like. We didn't know if there were more Germans ahead of us or where the Allies were or anything, but we thought it's now or never. So we took off and didn't run into any Germans at all. We came to a big field of high grass and started across. We could see little figures at the end of the field flitting among the trees. They looked like they were coming our way and we thought they probably were not Germans. We really didn't know the best way to go. Do you let them overrun you? Do you hide or identify yourself as American or what? We decided to just walk right into them so we just marched across the field and, probably lucky for, us we ran into the most professional soldiers in the world, Indian Gurkhas. Little bitty guys but very fierce and very courageous. We got in among them and kept saying 'Americans, Americans'. They weren't concerned and nobody raised a gun but they didn't understand us; also they didn't do anything and weren't threatening. All of a sudden a British first lieutentant appeared and we were home free. We told him our story and that we had intelligence that a large German reinforcement was on the way. He turned around and told the Gurkhas about that and they all broke into big grins. They thought that was great.

"We didn't want to get caught in the middle of a fight so we told the lieutenant that we'd like to get to the rear as soon as we could. He directed us to company headquarters where there were a lot of Gurkhas and Sikhs sitting around having lunch. From company we went along to brigade headquarters. We were there a while, long enough to see about a hundred German prisoners brought in by the Gurkhas. We then went to division headquarters and they took us to an American division where we spent time with the intelligence people who wanted to know how we evaded where we'd been and so on. Then they took us back to our outfits. Lou to the 82nd and me to the 1st. Most of my friends were gone, either shot down or gone home. My operations officer was now the squadron commander and a major (Lee Wiseman).

"Walt Morrison and I wanted to stay with the 1st but that was against policy. We went through channels up to the Brigadier who was head of the Fifteenth Fighter Command all to no avail. I remember Morrison saying, 'Say, General, you're a general and you ought to be able to help us.' He explained the policy to the effect that in our situation if we were caught again by the Germans we could

Lt. Jim Dibble 94th, Mateur, KIA 9 Sept 1943. (Dibble)

be executed. So he said 'no, you guys gotta go home'. Before going however, we travelled around to the other bases and told them our experiences and what we could of evasion techniques. What to do and what not to do."

Dick Catledge's story was typical of several 1st Fighter Group evadees who made their way down much of the length of Italy in the winter and spring of 1943-44. Catledge was gone almost exactly nine months. In an interesting parallel Chuck Yeager was shot down over France in early 1944. He evaded with the help of the French partisans and returned to England and was told he was going home. He appealed just like Catledge and Morrison and won his case with results that we all know. The difference, Yeager wound up by having his case heard by General Eisenhower and it was left to Eisenhower to decide. He favored Yeager.

Missions during the summer of 1943 had the common although unannounced objective of laying a foundation for the invasion of Italy at Salerno by the United States Fifth Army commanded by General Mark Clark. The Salerno landing complemented the landing at Reggio di Calabria by the British 8th Army. The Germans, of course had concluded that Salerno was a likely target given the same facts that our Army commanders had. They just didn't know when. One of the German concerns was a United States penetration across to Foggia trapping German army units in the south. That, of course, did not happen. Such a crossing over the Apennines would have been difficult with no German army at all. Even in 1944 our tanks kept sliding off those twisty little mountain roads. As is usually the case, Salerno represented a compromise. The invasion could have taken place closer to Naples, i.e., north of the Sorrento peninsula but more distant for air coverage from Sicily. Salerno presented several military advantages among them better air coverage. It was still only some forty miles from Naples. Unfortunately many of those miles were mountainous.

The 1st Fighter Group assignment on August 30th seemed to be more of the same. An escort mission shepherding B-26s to the Aversa marshalling yards. Once again we were escorting the 319th and 320th Bomb Groups. The 1st formation, with Lieutenant James Dibble leading the 94th and group, caught up with the B-26s over the Mediterranean, assumed weaving position above the Marauders and continued on course for Italy. The B-26 bombing altitude was to be 12,000 feet. The groups crossed the coast of Italy and bombers tightened formation as they approached their initial point. The P-38s loosened somewhat for better maneuverability. All eyes were alerted for enemy fighters, and suddenly they appeared, approaching from the Naples area, high and fast in two waves.

1st Group flights immediately dropped belly tanks and broke into the approaching enemy fighters. From altitude the swirling dogfights descended nearly to the wave tops and seemed to grow in intensity. The expansion was real, additional German fighters had arrived from Foggia despite the pounding of a few days previous. The new arrivals dropped their belly tanks and dove on the outnumbered P-38s. As the fighter battle edged out to sea the B-26s completed their bomb runs dropped their noses and "GOT THE HELL OUT OF THERE!" Remarkably, not a single B-26 was lost.

The Presidential Citation awarded to the group describes the battle as follows:

> "Crossing the Italian coast their formation of forty-four P-38s was intercepted by approximately seventy-five highly aggressive and persistent enemy aircraft. Alone and unaided the 1st Fighter Group engaged them, beating off wave after wave of enemy planes attempting to pierce the fighter defenses and break up the bomber formation. These courageous pilots fought a brilliant defensive aerial battle, destroying eight, probably destroying three and damaging three enemy fighters while our own losses totalled thirteen missing. Through their highly effective cover, the bomber formation was able to complete a highly successful bombing run unmolested... and returned to base without the loss of a single bomber."

Of the thirteen missing pilots the 94th lost five; Lieutenants John Cram, Ralph Peck, Charles Woodward, David Parlett and Harold Rigney. Hal Rigney and David Parlett were captured, escaped and returned to the squadron in the summer of 1944. Jack Pettus was one of the more fortunate ones. Flying 'tail-end charlie', he was the only #4 man to make it back that day. From the 71st, Lieutenants John Willey, Clarence Southard and Walter Morrison were missing. Lieutenant Ralph Turrentine was presumed killed in a mid-air collision with German fighters. (There was some later speculation that Willey, the 71st leader that day, was the one killed in the mid-air, however this is not mentioned in the squadron mission report). The 27th squadron lost Lieutenants Harry Warmker, Clifford Randol, Jerome Weinberg and F/O Donald Deisenroth. This was the greatest loss the 1st Fighter Group ever suffered in a single day.

Bob Vrilakas, the leader of Blue Flight 94th squadron had his plane riddled with machine gun and cannon fire and barely made it to Sicily. He describes the battle from his vantage point:

"Going in with the bombers we encountered heavy flak then several fighters. The bombers made their run and we were on our way out when someone called a new batch of bogies, estimate 70+ high. I looked up to see them drop belly tanks and dive into the middle of us. A real melee ensued in which our flights were broken up and much of our squadron integrity disappeared. There were P-38s and Me.109s everywhere. I saw several P-38s on fire and one (Rigney) bellying into the water offshore.

"We had expended most of our reserve gas in the first encounter so the squadron leader that day (Lieutenant Dibble) gave the order to work toward the Bay of Naples and reform as best we could. About the same time a P-38 from the 27th went by me toward the Bay with an Me.109 closing in on his tail. I was alone and broke into the 109. I got hits on him and he broke off immediately, but my maneuver had taken me back into a beehive of enemy fighters. The squadron by then was loosely formed up and headed out to sea.

"It was obvious the 109 pilots were choosing lots to see who would get the honor and that there was no way I could best their numbers or altitude advantage. I took the only real choice available which was to firewall it and head for the deck out over the Bay. Almost immediately tracers went by; they were actually converging in front of me and I looked back into a spiral painted spinner not more than a hundred yards behind me. As I leveled off, at what seemed like a few inches off the water, he started hitting me. It sounded like hail on a tin roof with occasional explosions as his cannon shells struck. I kicked left rudder to skid and noticed his tracers going over the end of my left wing. As he worked back I tried a skid the other way. It reduced his effectiveness but he still got quite a few hits as he worked back through each time. I was so low on the deck that he no doubt had a problem avoiding the water plus my prop wash. At times his bullets raised plumes of water ahead of me indicating he was trying to shoot from an elevated position and would over-lead me.

"I was sure I was either going to lose an engine or catch fire and was considering how to handle that when I noticed I was rapidly overtaking another 109 directly ahead and above me. He was chasing the rest of the Lightnings up ahead. If I went under him then there would be two of them to contend with. If I pulled up and eliminated the skid to fire at him the guy behind me would have a sitting duck. I pulled up, got a good bead on him and fired all four 50s and the 20mm cannon. He started to smoke and went into a gentle turn to the right. Almost simultaneously there was a crash as the guy behind me gave me a large burst. One of his bullets creased the canopy bar beside my head which knocked out the whole rear of the canopy. Plexiglas flew all over the cockpit-even into my mouth.

"I dove for the deck again and for some reason experienced no more shots from my pursuer. I then found that my right engine oil temperature was rising rapidly and the pressure was dropping almost to zero. I feathered the right engine and watched the left with no little anxiety but despite gaping cannon holes in both wings and numerous machine gun holes in the engine nacelles and wings, the left engine continued to function fine. According to the gauges there was no rapid fuel loss. I then made contact with the squadron by radio and they did a turn while I caught up. Three of the 94th pilots, Dibble, Pettus and Anderson stayed with me while the rest proceeded back to home base at Mateur. We headed for Sicily, the nearest friendly landfall. After what seemed like endless time we arrived over Sicily and proceeded to the nearest airfield. It was under construction but there was enough of the PSP (Pierced Steel Planking) runway to land on so I set up an approach. Just as I reached the end of the runway an Italian truck being used for construction drove directly across in front of me. I managed to add enough power to get over him but decided against a go-around not knowing what the condition of the hydraulic system was. The landing went OK and I had enough speed left to taxi over to some tents being used by the Army Combat Engineers. As I shut the engine down it caught fire which was quickly extinguished by an awaiting fireman.

"The plane looked like a sieve. There were sixty to eighty .30 caliber holes plus four or five cannon hits that in one case flared up a large piece of wing skin. Even the props had bullet holes in them. Dibble, Pettus and Anderson after seeing me down safely, buzzed the field and continued on to home base. I stayed at the field that night then made my way back to Mateur the next day. (The P-38 did not.)

"I'm sure I got the 109 that I fired on ahead of me out over the water and theorize that the reason the 109 behind me stopped firing was that either he ran out of ammo and/or went to the aid of his comrade."

David Parlett's ship was set afire by attacking Me.109s. He bailed out, suffering some burns in the process and managed to inflate and climb into his dinghy with assorted burns literally from head to toe. His hair was burned off, one knee hurt and an ankle burned. He was finally picked up by an Italian patrol boat and brought ashore. He ended up in the same prison with Richard Catledge and their experiences ran in parallel until they finally reached Allied territory some nine months later.

Another story dealing with this mission concerns Hal Rigney, one of the 94th pilots shot down that day. Hal's case was different. He was shot down by a P-38-one with black crosses on it. Rigney was flying the number 4 position or 'tail end charlie' in his flight. He noticed the P-38 behind him but the whole squadron was so engaged that it was nearly every man for himself and one P-38 out of position was not surprising. That is until it started firing. Even then Hal's first thought was that someone had hit the gun button instead of the radio. Both were on the wheel. At any rate Rigney wound up on fire. He made preparations to bail out. With all the smoke and flame he couldn't determine his altitude and wound up bellying into the ocean some miles off the Isle of Capri. He was plucked out of the water and became a prisoner of war until he escaped. In his travels behind enemy lines and after Italy had capitulated he had occasion to talk with some Italian fighter pilots and mentioned to them that he was shot down by what he believed to be a P-38 with German markings. They confirmed that a P-38 had indeed been captured and used by one of their pilots against American planes both fighters and bombers.

Coincidentally, a letter from John Upton, a bombardier with the 301st Bomb Group described an attack by a P-38 which he witnessed. The P-38 side-slipped into the B-17 formation as they were withdrawing from a raid over Italy in the summer of 1943. The fighter was on single engine. After a time the pilot started his feathered engine pulled out and made a firing pass through the B-17 formation. He riddled several B-17s and downed at least one according to Upton. Some weeks later another P-38 approached the same B-17 group and was promptly shot to pieces by every gun the B-17s could bring to bear. The second P-38 was legitimate, much to the chagrin of the B-17 crews. They did try to make amends by throwing a get together with booze all round for the 1st Group. However, the second P-38 did not belong to the 1st.

Regarding the heavy Luftwaffe opposition of August 30th, one could almost say we asked for it, we being the Twelfth Air Force. The 1st Fighter Group just happened to draw the short straw. For days we had been bombing the same general area south of Naples using the same medium bombers at about the same time even from the same altitude. The wonder is (on August 30th) that the B-26s got out of there with no losses. It's interesting to note that the 82nd ran into the same kind of hornet's nest in the same (Naples) area just a few days later (September 2nd). Results weren't quite as bad, but the 82nd had more than a few anxious moments. The other strategic difficulty that remained for much of the war was the defensive posture of bomber escort. Be it medium bombers in 1943 or heavies in 1944, the escorting fighters were positioned over the bomber stream always constrained to fight a defensive action. The enemy could pick their spots; meanwhile the escort fighters continued their weave pattern over the bombers with a foray by flight or squadron depending on the enemy thrusts. We were usually on the defensive, rarely in an offensive role. This is borne out in the language of the citation and is one reason the low-level Foggia raid did so much for morale. That raid had been totally offensive, a great success and losses were as low as one could reasonably expect.

The August 30 mission did lead to an examination of tactics. The twelve ship three flight finger-four formation had evolved early in 1943 and had remained unchanged ever since. The squadron formations were now modified to 4 flights of the same finger-four configuration. This simplified the weave pattern but required more P-38s and more pilots, neither of which were available in any abundance. In fact the shortage of aircraft would grow more and more acute as the buildup of the Eighth Air Force in England siphoned more new aircraft, particularly the long-range Lightning.

Nonetheless, the war continued and the invasion of Italy moved closer and closer, and the Italians grew less and less enthusiastic about the war in general and Germany in particular. This feeling on the part of the Italians had manifested since Mussolini was deposed on July 25, 1943. His successor, Marshall Badoglio, exerted less and less control. In late August, as Archie Jackson remembers it, the field at Mateur was alerted to the possibility of incoming (defecting) Italian planes. (Italy surrendered to the Allies on September 3rd.) Orders specified lighting the runway at night which made everyone nervous in view of the sporadic Ju. 88 raids. These orders were carried out by Major Peddie who assumed temporary command of the 1st Fighter Group early in September relieving Colonel Garman who was reassigned to the United States. Garman had long since completed his combat tour and had been a member of the group from Selfridge days. His permanent replacement, Lieutenant Colonel Robert Richard, would formally assume command on September 19th. At the same time Major Horace Hanes assumed command of the 71st Squadron.

On September 4th P-38 units strafed German airfields around Naples. Since this had to involve some of the Germans who had taken part in the August 30th battle the 1st went at it with particular zeal. Smokey Vrilakas caught two 109s on the ground and flamed both.

Chow time, 27th Squadron, Sicily, 1943. (Bob Share)

Following an escort mission on September 6th, aircraft of the group did not return to Mateur but landed at Dittaino on the East coast of Sicily not far from Catania and south of Mt. Etna. The raids on the 5th and 6th had been B-26 bombing missions against Grazzianise airfield. German fighters attacked in "yo-yo" fashion, i.e., dive and zoom or sometimes a diving pass only. This type of maneuver was very difficult to deal with. Once an Me.109 got into a dive Lightning pilots were limited in what they could do because of P-38 compressibility problems.

Over the next day or two the air echelon assembled in Dittaino. This location positioned the group that much closer to Italy in order to support the Salerno invasion. The Sicilian stay would not be long. The Salerno landings, Operation AVALANCHE, took place on September 9, 1943. The stay at Dittaino was short but the days and nights were busy. The distance to Salerno was about 200 miles which allowed plenty of time over the beach even carrying one bomb in place of one belly tank. The group patrol areas were code named Pears, Peaches and Apples and comprised the area from the Sorrento peninsula to Agripoli on the southern end of the Gulf of Salerno. Each squadron flew two or three missions of one or two flights each.

On one late afternoon mission James Dibble of the 94th was flight leader. Stan Wojcik, a transferred P-40 pilot, was flying Red #2 on his first mission. The flight was requested to strafe German reinforcements in a valley some thirty miles inland. As the four P-38s made their strafing pass they were met with a hail of anti-aircraft fire from below, both sides and even from above (shades of Kasserine Pass). Only two ships in the flight made it back to Dittaino. Dibble, the flight leader, was killed and Stan Wojcik, his wingman, bailed out of his burning ship and got burned himself particularly around the face and eyes. Captured, Stan was taken north via ambulance and train with various stops to treat his burns and wounds, one being a Catholic hospital. Stan was Catholic, and the nuns did the best they could given their severe limitations. In addition to his burns Stan was beset with a siege of abscesses and jaundice. He spent a good bit of his hospital time looking after a fellow prisoner, a badly burned B-25 pilot. After several months in German hospitals he was finally assigned to Stalag Luft 1 on the Baltic Sea. Russian troops liberated the POW camp in early 1945.

On a different mission on the same day Lieutenant Frederick Messmore a Cragmore pilot was killed when his P-38 exploded on take-off at Dittaino.

Ten missions were flown on the 10th of September. Lieutenant Stanley Stuber of the 71st was attacked from the rear and forced to bail out over Capaccio. F/O Joseph Boyd also went down on this mission, both pilots returned to the squadron although Boyd spent some 18 hours floating in his life jacket before being picked up by a British launch. Joe Boyd made it back - just barely - and his story is a fascinating one.

He was orphaned at age four and raised in an orphanage. Turned loose at age eighteen with $10 in his pocket, he joined the Army. After Pearl Harbor he volunteered for the enlisted pilots program and became a flying sergeant Class of 42-G. Most flying sergeants at that time were destined for the troop carrier or transport command and Joe was no exception. After a stint dropping paratroopers from C-47's at Fort Benning he was assigned to fly a C-47 to Egypt. Stationed at a field near the Suez Canal he ferried supplies to the Allied armies in the Libyan desert. Flying time accumulated rapidly - 150 hours or more a month.

Flying C-47's was not Joe Boyd's cup of tea and he put in for transfer to Lightnings on a monthly basis and was routinely turned down. He was finally told (by an orderly room sergeant) to apply directly to the Commanding General Twelfth Air Force rather than simply to his squadron commander. This he did and in due time received a response from General Cannon approving his transfer with the comment that the General had seen pilots try to avoid combat in one way or another but not too many requesting it.

By then a flight officer, Boyd hitch hiked his way to Casablanca, ostensibly to a school but actually to an assignment center, where he was checked out in a P-38 which had seen better days. Two days later with four and one/half hours in a P-38 he was assigned to the 1st Fighter Group, 71st Squadron. On paper Joe Boyd's credentials may have looked reasonable with over 1600 hours in C-47s, so off he went on his first mission flying the usual Red #2 position headed for the Naples area. In reality he could hardly have been more poorly prepared. He had no gunnery experience, in fact did not know how to turn on the gun sight, nor could he fly formation and this latter nearly did him in. First he would fall behind, then accelerate too fast and have to break out of formation to avoid overrunning the squadron leader. All of this used up his gasoline at a great rate; thirty minutes from base on the return trip he had an estimated ten minutes of gas left. Big Fence gave him a heading to a little used grass field where he managed to get down deadstick. Finally returning to the squadron the CO decided he needed more training particularly in formation flying.

On his sixth mission, this time covering the beachhead at Salerno his flight was jumped by Me.109s and F/O Boyd suddenly found himself in serious difficulties. His left engine was shot out together with his entire electrical system. With most of his gauges inoperative he had no way of knowing how much gas he had and no means of transferring gas from his dead engine side to the right engine. With the setting sun on his right side he set out for where he thought Sicily should be and found it, or at least Mt. Etna. Flying just off shore looking for a place to land or belly in, his good engine ran out of gas and Joe ran out of options. He ditched just off shore, so he thought. Getting out of the cockpit he tore his dinghy but inflated his Mae West and took stock. Shore wasn't more than a 1,000 yards away and he thought he could swim that distance easily enough but first he took off his shoes thinking that would make swimming easier, not stopping to realize he was swimming toward a volcanic rock beach.

He soon had more to worry about than feeding his shoes to the fish. By noon the next day he was treading water far out at sea and could barely make out the tip of Mt. Etna. The currents were definitely working against him. Luckily a British PT boat came by looking for one of their own pilots thought to be down in the area. They threw over a rope ladder which he was too weak to climb. After finally fishing Boyd out of the water the British discovered he was a Yank. Joe thought for just a minute they might throw him back.

After four days in a field hospital he returned to the 71st. Leaving the hospital the first person he ran into was an infantry officer. Boyd had no insignia, no wings (British souvenirs) barefoot, no hat, shirt sleeves rolled up and pants that had eighteen hours washing time in salt water. He started to ask the Lieutenant where he could get some shoes and instead got a royal dressing down. Every time Joe tried to interrupt he would only succeed in angering the Lieutenant again. After Boyd finally explained his problem, the infantryman stared, apologized and ended up bringing the bedraggled flyer back to the 71st in his Jeep.

Joe Boyd completed his missions - still a flight officer - and rather than be rotated back to the States, was assigned to ferry new P-38s to the group from Africa. In the interim, he says, Colonel Richard, the Group Commander, OK'd a commission. Eighteen days after being appointed a Second Lieutenant Joe Boyd was promoted to First Lieutenant at which point he headed for the USA. Boyd may hold claim to a record for shortest time in grade for a Second Lieutenant.

While the pilots were busy flying several missions each day the ground crews were hampered by a lack of equipment, and at Dittaino it was once again back to the backbreaking job of gassing the planes by hand from five gallon gas cans - the joys of the hardworking air echelon.

On September 11th Lieutenant Marion Phillips was lost. The Catania operation required some night flying, both very early morning take-offs and night training flights. Taking off and assembling in the dark is no mean feat. Phillips took off on an early morning mission, called in to report some trouble and simply disappeared without trace. Possibly the victim of vertigo over the ocean, no trace of a crash was found on land.

As an example of intense Salerno operations, four Cragmore missions were recorded on September 11th, the first up at 0455 hours to Pears area for beach area patrol. In mid-morning eleven P-38s, including four from each squadron, dive-bombed and strafed in southern Italy. At 1115 hours eleven Cragmores patrolled over Peaches area. Twelve 27th squadron P-38s departed Dittaino at 1540 hours assigned to the Capri-Salerno area. They were redirected to the Pears area, saw no enemy air activity but reported large numbers of A-36s, P-40s and Spitfire aircraft in the patrol zone. At 1650 hours P-38s from all three squadrons patrolled between Capri and Salerno. This group intercepted some twelve to fifteen enemy aircraft and Lieutenant Rodney Fisher of the 71st shot down an FW 190 and two Me.109s to bring his total of victories to five. The Focke Wulf exploded, one Me.109 crashed into a mountain and the second went into the water. The Lightnings returned at 1930 hours to complete the ninth mission of the day. And that was a typical day at Dittaino for each of the squadrons.

James Alford, in his history of the group, described two incidents that well illustrate the hazards of night flying. At 0530 one morning Petdog Red flight #3 and #4 (Alford was #4) were waiting for Red leader and Red #2 to complete their take-off. Suddenly a ball of fire erupted at the end of the runway. Lieutenant

Warren Eickman flying as Red #2 had evidently hit propwash, lost control and crashed sideways with a full load of gasoline and ammunition. The plane immediately burst into flames, however Eickman leaped out and raced clear of the inferno. Meanwhile Red 3 and 4 could only wait, stare at the flames and wonder what had happened. They were immediately waved off to continue the mission. In an even more unlikely event Lieutenant Frank Gerry, in another night flight, lost track of his flight leader. Looking around he spotted what he thought were P-38 wing navigation lights. Closing, he discovered too late that the lights were marker beacon lights for a British air field in the vicinity. Gerry hit the ground at nearly 300 MPH for the fastest belly landing in history. He finally skidded to a halt as his P-38, shredding parts, burst into flames and ammunition began exploding. Unbelievably, Gerry escaped more or less in one piece but suffered injuries which hospitalized him until mid-November.

The Germans counterattacked strongly against the Salerno landing on September 12th and the group was kept busy with repeated dive-bombing and strafing attacks against elusive and dangerous ground targets. Bob Van Sice of the 94th recalls being vectored inland by the British command ship to, "strafe some trucks". The trucks turned out to be units of a German Panzer division. Bob called it an exciting and terrifying experience. By the 12th of September the German's had managed to bring to the Salerno beachhead elements of four panzer divisions. These led the counterattack that very nearly succeeded. The Germans drove to within 1,000 yards of the sea and came very close to splitting the Allied forces in two. It took the combined efforts of the ground, naval and air forces to save the Allies being driven into the sea. By the night of the 15th the attack had been stopped. Supporting naval gunfire and air attacks had caused such heavy losses that the Germans had no choice except to begin withdrawing.

Frank Lawson, a Petdog pilot, makes a point and counterpoint to the Salerno operation:

"In late August 1943 along with five other P-40 pilots I found myself assigned to the 27th Fighter Squadron even though I had never seen a P-38 except in pictures. To make things even more bizarre I probably had less than 250 hours total flying time. After a "check out" and three or four flights I was assigned to the first flight to cover the beachhead at Salerno on the morning of the invasion. After a late start because the Squadron Commander had to show me the light switches, I was able to take off directly over my element leader who had crashed. One of the other 'new guys' also crashed into the side of Mt. Etna and I wound up all alone heading north. Confused in the dark and devoutly frightened, I continued until flak bursts convinced me that I had gone too far. Indeed, I was right over Naples.

"Turning out to sea, I was just in time to see the entire invasion fleet open fire. There must have been 300 ships out there all firing at once. Believe me, I was impressed.

"After daylight, I was able to join up with one other P-38 and complete the patrol. No Germans showed up but one British Seafire from a carrier almost got into it because we had been told that we were the only Allied fighters that would be in the area and any single engine aircraft was, naturally, hostile.

"Twenty-five years later I went to war in South East Asia as an F-105 pilot. The training lasted six months, including three survival schools, air-to-air and air-to-ground firing, hundreds of practice bombs and some 1,000 pounders, day and night in-flight refuelings and everything that one could possibly do to become completely familiar with the F-105.

"The irony of these two experiences; we won the first war and lost the last one."

The German counter offensive was broken and by September 16th the Fifth Army was well off the Salerno beaches and had started its long, bloody, slogging advance up the mountainous Italian peninsula. On the 18th the 1st Fighter Group returned to Mateur after twelve hectic and dangerous days. The group had flown a total of eighty-seven missions in support of the Salerno beachhead. In retrospect the Catania experience represented yet another example of remarkable coping which became the norm. The 1st dropped men and planes into a wheat field in Sicily, overlaid with steel matting, organized the logistical equipment and became a functioning, effective fighting unit with nothing more than manpower to pump gas, move planes, load bombs (sometimes until midnight) and eat, sleep, wash. They saw P-38s shot to pieces, pilots killed-or exhausted, and yet supported a successful invasion against an entrenched, determined enemy. The return to Mateur earned the group a well deserved two week stand down from operational flying.

On September 22nd the group provided escort for units of the Regia Aeronautica following Italy's

Sandstorm, Gambut, Libya. (Lee)

surrender. Over 200 Italian aircraft arrived at various bases in North Africa. Naples was taken by the Americans on October 1,1943. The surrender of Italy had several significant consequences. Among these was the elimination of thirty Italian divisions from the Balkans forcing the Germans to take over policing that area, an unenviable task. The surrender also removed the Italian naval fleet as a potential problem and last, but not least, simplified and hastened the takeover of Adriatic coast airfields for use by Allied long-range bombers and fighters.

On a dive-bombing mission of October 3rd the 71st lost three pilots. Lieutenants Hobart Vincent and James Conner collided in mid-air. Vincent's plane spun in and Conner went down in flames. Ralph Parrish, also of the 71st, failed to return from this mission. The remainder of the month saw the complexion of the group begin to change once again with an exodus of veteran pilots back to the States. The returnees included Major Otto Wellensiek and Lieutenants Sokol, Niles, Anderson, Newsome, Smith, Hurst, Helm, Fisher, Chapman, Schreffler, Nygren, Russell, Sentes, Bower, Pace and Opdahl. Major Horace Hanes formally assumed command of the 71st Squadron, Captain Gilbert Butler took over the 27th and First Lieutenant James Hagenback became C.O. of the 94th Squadron.

Early in October, in a surprise move, the Germans invaded the island of Cos in the Dodecanese. This particularly bothered British troops in the same islands who found themselves out on the end of a limb a good distance from their base of supplies. The decision was made to evacuate the Dodecanese island chain (located in the eastern Mediterranean). To accomplish this the British evacuation fleet would need air cover to protect the ships from German aircraft based in Greece. For this reason the 1st Fighter Group and the 14th Group found themselves enroute to Gambut, Libya, (an airfield near Tobruk) on the morning of October 5th. The P-38s flew out across the Gulf of Gabes then east across the desert. As the afternoon shadows lengthened one sand dune began to look just like every other sand dune. The formation finally spotted what looked like an airfield and landed followed by the air support echelon in C-47s. The field wasn't Gambut but it wasn't far off. After suffering through the usual cold desert night, everyone flew the few minutes to Gambut the next morning. The pilots by this time knew how to prepare for these itinerant travels. Usually every nook and cranny on the P-38s would be crammed with extra blankets etc. Unfortunately there wasn't much to be done about food. K-rations could be packed but short of imminent starvation they were a poor choice.

The following several days were spent alternately providing air cover for the British fleet steaming near the Dodecanese and fighting sand storms. For the most part there were few enemy encounters, although the fleet for days was within easy range of German aircraft in Greece. The exception was the "turkey shoot" which befell the 14th. The Germans, evidently misreading their radar, saw the 1st Group planes departing and dispatched a flock of Stuka dive-bombers. Unbeknownst to the Germans the 14th had replaced the 1st. The 14th pilots couldn't believe their eyes but fell upon the Stukas and picked them off like flies.

The principal hazard at Gambut was sand. Returning from their cover mission on October 9th Dick Lee's flight ran into a raging sandstorm. Dick took his (White) flight 125 miles to the east and landed at a Greek field run by the British, got treated royally, then returned to Gambut. Once the fleet had moved out of range of

Lts. George Parker and John Lion, 71st Squadron, Gambut, leaning on one of the many wrecked German aircraft that littered the airdrome. (McCann)

94th pilots Gambut L - R Lee, Hoffman, Darnell, Hanton, Vrilakas, Howard, Brown. (Lee)

air attack the P-38s left Gambut and its daily sandstorm and returned to Mateur to face another kind of storm. In late October the cold weather rains were threatening to turn what was a dry lake bed into a sea of mud.

On the way back from Gambut Ray Ainsey managed to enjoy a side trip to Alexandria. "At the American Red Cross had fried chicken, a shower, beds with sheets and, just down the street, Scotch with soda and ice. Quite a shock compared with what went before- and after. On October 15th it was back to Mateur - too muddy to land - on to Bizerte and back to C-rations."

About this time, in what would turn out to be a classic in bad timing, but a very good idea, the group C.O. decided that there should be clubs for both the officers and the enlisted men. Using the two materials most readily available, packing crates and Arab labor,

Officer's Club, Mateur. (Lenhart)

the projects were started and completed. Bob Vrilakas describes the club as looking pretty good with a bar, dance floor, tables and chairs and a parachute draped on the ceiling. The bar was complete with a sign that read, "Candy is dandy but liquor is quicker," this in a club that had neither. There was, of course, the liquor ration of one ounce of rye whiskey per mission. That was usually saved up then each squadron would be issued the appropriate number of bottles, usually something like Old Overholt, dubbed Old Overalls. For the club's grand opening one of the more persuasive pilots was sent to the hospital in Bizerte to invite the nurses to a dance. They came, looking ravishingly beautiful; someone located some bourbon to supplement the Old Overalls, and a good time was had by all dancing to the music of a local orchestra. As the nurses were loaded back into their ambulances for the return trip to Bizerte it was the consensus that this was the first of many such affairs. Unfortunately, not too many days in the future, an air echelon section of the group packed up and moved to Djedieda leaving all the new construction to the local Arabs.

The construction of clubs was a particularly good idea for the enlisted personnel. The pilots had their fifty mission target but there was no such goal for the non-coms or the non-flying officers. For them it was day after day of tent living and C-ration dining, beset by flies (screening can only do so much), scorpions and then came the fall rains and the resulting Tunisian mud. It was not a fun time and there seemed to be no end in sight. People railed and bitched against the system, the Army, the chow, the mud, Africa, the mail system; in short everything but their own outfit. Bitching about the Army usually excluded one's own squadron.

On October 21st, escorting B-26s to targets in Italy, P-38s of all three squadrons were jumped by about fifteen enemy aircraft. We accounted for five 109s and one FW 190 but lost three Cragmore pilots; Lieutenants Junior Downing and John Lion together with F/O Spaulding did not return. A couple of days later in a repeat performance the 27th tangled with about twenty enemy aircraft and downed three of them plus several probables. Lieutenant Dwight Terry was among the missing. On this mission the 71st and 94th, flying ahead with the bombers, heard the 27th call in the enemy aircraft but could not locate the fight.

Targets were becoming more distant and the 1st Group's stay in Africa was drawing to a close. On October 29th, just prior to the air echelon move to Djedeida, much of the ground contingent began to move by boat (LST's) to Cagliari, Sardinia. They landed on November 1st and began getting established for what would turn out to be a short stay.

On October 31, 1943 the convoy with the ground echelon of the 1st Fighter Group docked at Cagliari, Sardinia. At least a part of the group was out of Africa nearly a year after landing near Oran. The 1st then embarked upon a month of dual operations with the air echelons still in Africa and the ground echelon settling in around Monserrato air field. Quarters varied; some people were quartered in a villa of sorts near the airdrome, the 27th was located in two story stone house as headquarters with people camped in various buildings around the main house; all in all, quite pleasant. We were also quartered in what had once been an apartment house of sorts. It had a roof, rooms with windows but no heat. There was also a defunct spaghetti factory. At least everyone was undercover without resorting to pup tents. Cagliari itself was in ruins and nearly deserted. The local people around Caglairi seemed to be every bit as bad off as those in Africa. They crowd around the camp at mealtime looking for any kind of a handout. Our fellows were generous as usual and ended up feeding a good number, especially the kids many of whom seemed totally destitute.

On November 1st the Fifteenth Air Force was formed the better to separate tactical from strategic objectives. The new air force, headed by General Jimmy Doolittle, was headquartered in Tunis but would soon move to Bari, Italy. The Twelfth Air Force remained tactical and the Fifteenth took over strategic operations. The new Fifteenth included four fighter groups - three P-38 groups and one P-47, (the 325th Fighter Group); six heavy bomb groups and, at that time, five medium bomb groups. In a few months the medium bomber groups rejoined the Twelfth Air Force, leaving only fighters and heavy bomb groups in the Fifteenth. The new air force was formed over considerable objections from the ETO, mostly British, who felt that its formation would diminish the Combined Bomber Offensive and ultimately jeopardize Operation OVERLORD.[2] The Americans felt that the CBO could be better fought from both England, by the Eighth and from Italy by the Fifteenth Air Forces.

The new air force had a struggle on its hands for the first four months with its four fighter groups at about half strength. The U.S. Air Force history by Craven and Cate states that the theater was short about 200 fighters at the end of 1943.

1. All three P-38 groups actually took part in this raid. The 14th flew as top cover for a high level B-17 followup raid later in the day to the same general target area.
2. Code name for the invasion of northern Europe.

10

OUT OF AFRICA

On November 6, 1944 the 1st Group finally got out of the mud but only partway out of Africa. The air echelon moved to Djedeida, some 60 miles from Tunis, an airfield with asphalt runways and taxi strips. (This about three weeks after the clubs were completed at Mateur.) The P-38s shared Djedeida with some British Wellingtons but it was certainly an improvement over Mateur. Djedeida would be a short stay. The rest of the squadrons were already settling in at Monserrato near Caglairi, that is they were at first. Originally about 50 ground personnel moved to Djedeida "for a week". The week turned into a month during which time about 100 of the ground people were air lifted back to Djedeida from Cagliari.

Not long after arriving at Djedeida the 71st lost Lieutenant Paul Binsley without explanation, not the first time such a thing happened and certainly not the last. Binsley took off on a test flight and disappeared. Test flights were usually very prosaic affairs. Slow timing new engines, checking out mechanical repairs of various kinds which couldn't be done by simply running up the engines on the ground. They certainly didn't involve high speed dives or the kind of high risk flying sometimes associated with the term test flight. In all probability Binsley lost his horizon and flew into the sea or met with some similar mishap.

The situation at Djedeida was not improved by the group being "bumped" by the RAF Wellingtons. The P-38s had to be taxied about a half mile on muddy roads and taxi strips which led to some accidents. To compound the problem the pilots sent their cold weather flying clothes ahead to Sardinia and were caught short with the few clothes on hand. On one B-24 escort mission at least two pilots developed serious cases of frostbite.

The stay at Djedeida was probably one of the most miserable of the war. If not the worst it ranked right up there. The beginning of winter brought a rainy cold that penetrated. The tents had no floors and there were no stoves and with few exceptions no winter gear. The food was C rations out of cans, lumped together and heated. Even the cigarette situation went to pot. In place of the usual Camels, Lucky Strikes and Chesterfields some very second rate brands appeared. Alford's history recounts that one pilot, confronted with a pack of Wings, observed it was surely an indication we were not winning this war.

Alford also describes a masterpiece of foraging by the 27th Squadron flight surgeon, Captain Archibald Adams. "Several weeks previously an ammunition dump had exploded and set fire to a part of a Quartermaster installation nearby. Johnny-on-the-spot very shortly thereafter was Doctor Adams in his omnipresent ambulance. He loaded this with some two dozen large cans of boned chicken. This delicacy was doled out to his pilots at Djedeida (only in the 27th of course) when darkness had fallen and only after all concerned were sworn to secrecy, silence and whispers."

From Djedeida the group flew one of its longest missions escorting B-24s to Athens with an outgoing fuel stop in Sicily and a return stop in Lecce, Italy where the 82nd Fighter Group was already based. Dick Lee's log on this mission shows 9:05 hours. Long escort missions continued one by one moving almost counterclockwise around the Mediterranean from Athens to Toulon, France. The Athens mission provided one close call as recalled by Smokey Vrilakas,[1] who was leading the 94th White flight on the way out when he noticed a single Me.109 barreling in on the rear of Dick Lee's Blue flight. Smokey called a repeated Springcap break right and nothing happened. He finally took White

L. T/Sgt. Pillard, R. Lt. Charlie McCann, 71st, Cagliari. (McCann)

flight around into the 109 which immediately broke off and headed for the deck. The squadron leader was upset about the break from formation until he heard what had happened. As is often the case, someone wasn't looking.

With the increasing emphasis on high altitude, long-range bomber escort the pilots became, by necessity, more oxygen conscious. This was underscored by the loss of at least a couple of pilots in what was suspected to be oxygen deficiency. Some were surprised to find their oxygen hoses dangling disconnected after a bout of turning and twisting their heads to keep track of enemy fighters. It was not simple to detect a loss of or insufficient oxygen flow, particularly if euphoria was beginning. Smokey Vrilakas says that his method of checking was to take off a glove and look at his fingernails. If they looked blue he needed oxygen. The supply was not unlimited and had to be managed until the squadron had let down to about 10,000 feet on the return trip. At that point one could take off the oxygen mask and light up a cigarette, and how good that tasted.

One testimonial to longevity was recorded in late November. On the 21st we escorted B-26's to bomb railroad bridges in Italy. Several enemy aircraft were engaged in the target area and two of our planes were damaged slightly. One of them was DEAR JOHN an old F model. This ship was flown originally from California to England during the BOLERO Mission, then to North Africa and had been on active duty ever since. The venerable Lightning had amassed over 600 combat hours including 41 dive-bombing missions and had accounted for ten enemy aircraft.

On November 29th all air echelons moved over to Monserrato, Sardinia. It was slightly more than a year since the 1st had touched down at Tafaraoui near Oran and finally everyone was out of Africa. Even so, the stay at Monserrato was brief. In just a few days the group moved again, this time to Italy and, once again as air and ground echelons. The group was leaving a trail of people behind in various hospitals with assorted ailments; some malaria but more dysentery and even some yellow jaundice.

The landing strip at Monserrato was hard packed dirt. It had earlier been used by units of the Italian Air Force in fact some of the Italian "Black Cat" Squadron[2] personnel were still on the field. While Monserrato was the first step out of Africa it was a very brief one in terms of missions flown. Dick Lee recorded two missions, both B-26 escort trips to Italy. On the second, December 8th, the target was completely overcast; so much so that no bombs were dropped. The time in Sardinia included Thanksgiving complete with turkey and football. The turkey dinner was complete with all the trimmings although there were some variants. Ray Ainsley noted in his diary that he and his friends enjoyed venison, wild boar and turkey in their quarters in Roses Spaghetti factory - that seems to qualify as a variant! 27th Fighter Squadron notes describe a Thanksgiving Day football game and this too qualifies as a variant if the description is accurate. The game was labeled a classic; officers vs enlisted men. The officers won 6 - 0 on a pass from Lieutenant Rafael to Lieutenant Wellner. So far so good but the lineups were strange. There were seven men per team, not six and not eleven. The officers included Wellner, Goebel, Rafael, Tibble, Maloney, Parsons and Ferrin. The EM's included Smith, Ladson, Meachem, Schneider, DiLorenzo, Hood and O'Meara.

During the stay in Sardinia the Luftwaffe was nearly as active as we were. On December 2nd German Ju. 88s and Heinkels raided Bari harbor. Flying from Greece they sank a total of seventeen ships. The bombs

DEAR JOHN, a P-38F, Ser. No. 41-7547. Pilot, Capt. Newbury, Crew: Brendon, Abbott, Emery and Coute. (Ilfrey)

German Stuka dive-bomber in Italian markings at Monserroto Airdrome. Working for the Allies after the Armistice, this aircraft was likely part of 121 Gruppo. (Lee)

hit two ammunition ships which exploded and accounted for several of the other sinkings. The raid destroyed 38,000 tons of cargo including hospital supplies and runway steel planking. It also killed about 1,000 and blew out windows in the new Fifteenth Air Force headquarters.

As the air echelon was coming in from Djedieda to Monserrator there was a needless accident. One of the 71st pilots, Lieutenant Russell Williams, landed on the dirt strip and blew a tire. The plane slowed nicely but veered to the right as the tire shredded, and at the edge of the runway slowly tipped up and over onto its back. A number of people, including some of the Italians of the Black Cat Squadron rushed out and started to raise the tail to extricate Williams but were ordered away by an officer who evidently feared a fire. As everyone stood around helplessly Williams was slowly choking on his helmet strap. When the order was finally given to raise the ship he was dead. The death of a pilot was certainly not unusual in those times, but the circumstances surrounding the death of Williams were depressing in the extreme and everyone felt it. The accident reminded Clyde Gabler of a similar incident at Ibsley in England when he and an airman extricated a Hurricane pilot likewise upside down and choking. In Gabler's case no recognition was warranted, "because," in the opinion of higher ups, "there was no fire."!!

With everyone in place at Monserrato the month of December did not start auspiciously. Escorting B-26s to targets in Italy, two 71st pilots Lieutenants George Parker and Edward Tanner collided in mid-air and became prisoners of war. Lieutenant Eldon Vondra, a Petdog pilot, was killed in a running twenty minute encounter with fifteen to twenty enemy aircraft.

On December 9, 1943 the 1st Fighter Group was on the move once again as an air echelon flew to Gioia del Colle, Italy. Coincident with the move Major Hanes transferred to group as Operations Officer and Captain Lee Wiseman took over as commander of the 71st Squadron. The balance of the group left Cagliari by boat on December 20th and docked in Naples on the 22nd. Details from the whole group were put to work in the Naples harbor doing everything from loading to sorting mail and guard duty. Not all was in vain, however. In true GI fashion they appropriated everything from food to Jeeps and took the booty along in the convoy to Foggia. Overall unloading of the ship was an unqualified mess. It took at least a week with the ship's crew managing to break our boxes, drop our trucks and generally confuse matters. The ground contingent arrived at Foggia #3 on the 29th after dark. It was, naturally, raining, cold and windy. Vance Tuttle's overall description of assorted living quarters was, "Hell, foxholes, pup tents, horse barns, pyramidal tents, spaghetti factory, back of trucks-you name it." Group supplies and equipment were in wretched shape from the confusion and rough handling aboard ship and shifting and dropping on the truck convoy to Foggia. Boxes open, pieces and lumber scattered behind, equipment soaked and everything cracked or broken.

In the Djedeida-Monserrato-Gioia del Colle-Foggia sequence of moves the air echelon operated independently of the total group to the considerable disadvantage of the former. The air echelon stay at Djedeida was a cold, miserable month. Everyone rejoined briefly at Monserrato, however, the air echelon stay at Gioia del Colle was equally bad in terms of living conditions, flying conditions and losses.

Harry McConnell describes an incident, amusing as it ended but descriptive of the chores involved in moving a squadron (in this case the 94th) from one location to another. It was more than simply packing your tent and sleeping bag. Harry describes his roundup of telephone wire and the ultimate epithet coined in the process. The epithet was aimed at one Hugh T. (High Tension) Jones.

"A close buddy, Ken Siemers, and I had been assigned the duty of retrieving the squadron telephone lines from Cagliari. The process was performed by the use of a Jeep with a large empty reel mounted aft. Starting at any given point, one simply cut the lines, laid in strands of two and reeled in.

"The lines we were to retrieve were laid on the ground (normal procedure) in a deep wide ditch, into which we could drive the Jeep. The duty may sound simple but the conditions were not. Many other lines had been laid in the ditch, and how this condition can develop I have never been able to understand. Our lines at intervals would pass over and under other lines, requiring that we frequently cut, untangle and splice. OK, Mission completed.

"Now we had to drive the Jeep out of the ditch to level ground. Unfortunately due to recent rain the Jeep could not get traction as we tried to pull out of the ditch. We hail a passer-by to drive the Jeep while Ken and I push from the rear. The volunteer driver guns the Jeep up the incline; Ken and I are splattered with mud, mud, mud! It so happens that Ken had borrowed H.T.'s raincoat to wear on this misendeavor. So much mud on the raincoat it looked like a suit of armor.

"When we get back to the building being used as "Communications Base" there is H. T. sitting on the workbench. H.T. looks at his mud coated raincoat draped on Ken and goes 'ape' - he's about to throw a tantrum! I'm pretty stressed about the experience Ken and I have had. H. T.'s reaction to the raincoat triggers me and suddenly I'm in H.T.'s face telling him just how 'chickenshit' I think he is!

"At that time the most complex radio tube in use was known as a 'duo-diode-pentode.' It was the ultimate tube of the day. How it came to mind I'll never know but I snarled in H. T.'s face, 'Jones you are duo-diode-pentode-chickenshit!' Jones reaction was a split second of mental suspension while he digested the words. The second was just the opposite of what everyone expected. He was soon roaring with laughter. He considered it the highest compliment and the funniest thing he had heard in many moons."

The move to Gioia, fortunately to be a brief one, provided a realistic introduction to Italy in the winter. On arrival there were no housing facilities worthy of the name, tents would not stay in place on the stony ground, and, being Italy, it rained. As soon as possible everyone

94th Sq. Mess area, Cagliari, Sardinia. (Lenhart)

71st Squadron crew chiefs, Monserrato Airdrome, October 1943. L-R Quinton, Puckett, Field, Pillard, Johnson, Simpson. (McCann)

moved across the road into some wooden frame buildings which someone insisted were, at one time, Italian chicken coops. The airfield had a partial concrete runway which quickly gave way to grass. The official record describes the situation at Gioia as follows: "There are several buildings and hangars on the field but they are occupied by the British who show no inclination toward cooperating with us by sharing. The weather is cold and damp and camp was set up and stoves are being made. The runway is just a grass strip with a number of low spots and gets very muddy after a little use." The field was located near a small Italian town about 40 miles south of Bari and inland in the hills on the eastern side of Italy.

Gioia and Italy were quite different from Africa. It was possible to get food in some small restaurants in town which, while they didn't have much, were a real improvement over powdered eggs, spam and above all C-rations. It's surprising how good fresh cooked potatoes can taste - just potatoes. However, one could not or should not get too particular. Smokey Vrilakas recalls, "... three or four of us enjoying bowls of hot soup in a restaurant. Mine was delicious until the head of a bird, beak and all popped to the surface."

Once established at Gioia we flew two escort missions to targets in Greece. The first to the Athens dock area and Kalamaki Airdrome was interesting for Cragmore pilot, James Ferguson. Turning back from the Athens area because of a shortage of oxygen he was attacked by three Mc. 202s.[3] He turned into them firing and observed his tracers hit the lead aircraft which spun nose over tail. He then fired at the second one, observed hits and then lost consciousness. He regained consciousness at 9,000 feet over the Southwest Coast of Greece from whence he returned to base.

The next mission from Gioia proved more costly. On December 20th, returning from an escort mission to Greece (Kalamaki and Eleusis), the weather was closing in over Italy. Lieutenants Harmer, Lipowicz and F/O Charles Brown all of the 94th, crashed into the hills and were killed. Lipowicz was alive when they reached him but died in the hospital in Taranto without regaining consciousness. The 27th Squadron lost Francis Mackle in the same manner. Lieutenant Devenney, a 94th pilot, flew through an orchard and had twigs from trees sticking from the tail of his plane. The crashes were a combination of bad weather, inexperience, low fuel and lack of any homing device at Gioia. To this list we should add rickety airplanes. No new aircraft had been received in the entire 1st Group since mid-summer. By the end of 1943 each squadron did well to put eight to ten planes in the air, and then only by a great deal of scavenging and cannibalizing by the mechanics. The ground crews were extremely good at their jobs but just had too little to work with particularly when isolated in an air echelon operation such as Gioia. In November Fifteenth Air Force estimated the P-38 groups at about half strength. Replacement aircraft had been steadily diverted to England to build up the Eighth Air Force.

One of the reasons for forming the Fifteenth Air Force had been the expected better winter operating conditions vis-a-vis the Eighth's operating conditions in England. Conversely the Italian weather proved so bad that only six of the Combined Bomber Offensive missions were flown in the last seven weeks of 1943. (CBO targets were German aircraft factories, ball bearing plants, oil and munitions as opposed to the more local communications, bridges, transport and the like). The planners obviously were not familiar with "sunny" Italy in the winter.

The serious problem of inadequate pilot training continued to plague the group. The P-38 was a complex airplane and could be unforgiving in the hands of a poorly trained pilot. Couple that with the dismal weather conditions often found in the Italian winter plus few navigational aids and the results could be deadly. As if to emphasize this Lieutenant Francis McGrath a Petdog pilot turned up missing over Italy on December 19th. He was last seen in a steep glide from 21,000 feet near San Martino Italy. No enemy action was involved and McGrath may have been a victim of oxygen starvation.

On December 21st the ground echelon departed Cagliari harbor bound for Naples. Unit records state that everyone was crowded aboard the **William J. Bryant** and, "there is not room to turn around including no quarters for sleeping except in the hatches. There is also no way to cook so C-rations will be the menu for the duration of the trip."

It turned out to be a harrowing Christmas. To begin with the ground echelon spent the day unloading their ship which finally docked in Naples harbor. On a mission escorting B-17s to the Udine under continuous overcast the 27th squadron was engaged by four to eight FW 190s. Separated from the main formation by the inconclusive scrap the 27th set out for home base. Meanwhile the B-17s were trying to climb over the overcast by doing 360 degree turns. As a result of all these maneuvers the rest of the group picked up the wrong wing of bombers and wound up escorting them to Bolzano, Italy. Ralph Kipper, the squadron leader, ran out of oxygen and had to return so Vrilakas took the lead. They were jumped by enemy fighters and had to drop belly tanks quite a ways short of the target. All of this added up to a real gas shortage. The official record says that all planes returned to base or landed at friendly fields. Bob Vrilakas' diary describes what landing at a friendly field can be like:

"Came back through bad weather between overcasts. Hines got very low on gas in fact we all did. Got a homing to the closest airport which was 30 miles behind our lines at Tremili. Was lucky in finding a hole in the overcast near an airfield so landed there. Couldn't get gassed in time to go on and weather was bad, so they put us up for the night. The 79th Fighter Group (P-40s) was stationed on the airfield, so they took us in and fed us a Christmas dinner which really hit the spot as we had not had lunch and had landed there at 3:00 PM. General Cannon flew a P-51 into the field and talked with us awhile about the mission. Swell bunch of fellows there they did everything to make us comfortable.

"Got up at daybreak the next day hoping to get the squadron back to our own base but it was raining and the Colonel there grounded us. We preflighted the planes in the AM and got them ready to go. Spent the afternoon talking with the fellows in the 79th. They had a small club in each squadron. We heard some new songs on their phonograph (Pistol Packin Mama and Paper Doll). They live a kind of rough life living so close to the front but get along pretty good considering the circumstances. It was the muddiest place I've ever seen. They live in tents and it was cold as hell. Went to a show in the evening and spent the rest of the night swapping tales about flying with the 79th pilots. Three RAF pilots wandered in about 9:00 PM . They were from an adjoining field and had been celebrating Christmas considerably. One of them had mud all over his face and clothing and went to sleep in a chair. His buddies explained that he had become a bit "troublesome" so they had to "discipline the bloke." Went to bed at 11:30, still raining.

"Awake early, weather considerably better. Got the squadron rounded up and by the time we got all the planes preflighted it was 10:30. Took off, gave the 79th a buzz job and headed for home. Weather not too sharp. Landed at field at 11:30 and it started to hail. It has been sleeting and snowing all afternoon-very cold."

In parallel with these adventures the ground echelon was struggling to get themselves organized at Foggia #3. They arrived late in the afternoon, the weather cold, wet and miserable and the field very muddy. As the record says "enough tents were set up to provide shelter for the personnel whose morale is far from high tonight." The only thing worthy of note was a Christmas dinner, turkey and the trimmings, served late but nonetheless served after the mess tent nearly blew away in a near gale.

On December 30th we had an interesting but unsuccessful rescue mission to Albania. Cecil Quesseth recalls, "A cargo plane carrying twelve nurses crash landed in Albania. We provided escort to a second cargo plane, a C-47 which would land, pick up the nurses from the local partisans and return. According to plan a British Wellington with paratroopers on board, would land first and hold the field. The P38s were to provide any strafing protection necessary to keep any German troops, if there were any in the area, at bay until the cargo plane took off. Apparently the Germans were alerted and were out in such force that the whole thing was called off. The Wellington put its wheels down hoping for a signal to land but nothing was forthcoming and,of course, the cargo plane couldn't land either. All of us returned to our respective bases, disappointed.

The nurses later escaped returning to Italy by boat. They had been en route by transport from Africa to Italy when fog and bad weather caused their plane to overshoot Italy entirely and they crash landed in Albania.

Most of the records simply describe this mission as "secret"; there is no mention of nurses but there's no denial either.

As 1943 came to a close the air echelon was operating out of a field shared reluctantly by an RAF Spitfire unit. The ground echelon was doing it's best to ready its quarters north of Foggia. By December 1943 the ground personnel had reason to gripe. To date no ground officer or enlisted man had been returned to the USA after 19 months of foreign service. Some had served in grade for over 20 months without any hope of promotion other than by transfer. Pilots rotated home after completing fifty missions-in some cases 150 combat hours. On the average it took four and one-half months to complete fifty missions. Not that it was easy, but with invasion coverage pilots flew two and sometimes three missions in a day. It should be noted that the four and one-half month figure was increased drastically during the winter and spring months. In a retrospective mood the group historian noted that in January 1942 the lack of replacement airplanes had been a serious problem. This was gradually alleviated during the summer months but had again become a problem at year's end. On a December 30, 1943 escort mission the group could muster only thirty aircraft or an average of ten per squadron.

Group administration also had its problems, the largest being miscellaneous reports and battle casualty reports. "Slowly the duplication of reports has been eliminated and reports required from subordinate units have been reduced. Battle casualty reports have been revised, rescinded and revised yet again causing duplication of efforts and administration. Revised Army regulations have been made available long after the effective date of the revision."

The group medical officer Captain Sposato also cast a look back at 1943. He joined the group in October and none of the medical staff predated May so the look was second hand and mildly positive. He did remark on the poor living and flying conditions at Biskra and Youks les Bains, the increased risk of malaria and dysentery with the coming of warm weather and the poor food at Mateur. He noted in particular that many of the pilots refused to eat the C-rations at Mateur, which combined with the swarms of insects made for grim days. In perhaps an unrealized indication of priorities Sposato noted that pyramidal tents were made available at Chateau d'un, "mostly to the officers, crew chiefs and cooks." The Captain concluded by stating that the year has generally been much better than anticipated. That may be damning with faint praise but someone must have anticipated a really lousy year.

The casualty record tabulates the following as of December 31, 1943:

Killed in action	**21**
Missing	**76**
Prisoner of war	**16**
Killed in Accidents	**6**
Wounded/injured	**16**
Escaped and returned	**7**

The following pilots had been credited with five or more victories:

	Destroyed	**Probable**	**Damaged**
Maj. Darrell Welsh	**5**	**0**	**3**
Lt. John Wolford	**5**	**0**	**1**
Lt. Daniel Kennedy	**5**	**0**	**2**
Lt. John MacKay	**5**	**1**	**0**
Capt. Meldrum Sears	**7**	**0**	**0**
Capt. Lee Wiseman	**5**	**0**	**0**
Lt. Rodney Fisher	**5**	**0**	**1**
Capt. Jack Ilfrey	**5.5**	**0**	**2**
Capt. Newell Roberts	**5**	**1**	**0**

Finally the year 1943 came to a close on a wet and muddy base where the group would be quartered for their longest stay of the war. The Salsola post would slowly take on some of the aspects of a permanent station until once again daunted by that ever constant enemy, mud. But that would be well in the future.

The three P-38 Groups moved to Italy at approximately the same time, i.e., late 1943 albeit by quite different routes. All three were positioned near Foggia by mid-January 1944.

As 1944 began the group lost Lieutenant Muffitt, a 94th pilot, on an escort mission to Sofia, Bulgaria. No one saw him go down and he did not survive.

The shortage of aircraft became strikingly evident on January 7th. On a B-17 and B-24 escort mission to Wiener-Neustadt outside Vienna, the entire group was able to put up a total of just twenty after aborts. Unfortunately, the 1st Fighter Group was all alone in decidedly unfriendly skies that day. The bombers had diverted to alternate targets, the Maribor Aircraft Factory and the Fiume Torpedo Factory. Under hazy, cloudy skies in the Wiener-Neustadt area the 94th and 71st, all sixteen of them, ran into some fifty to sixty German fighters in two separate waves.

Combat began at 20-22,000 feet and ranged down to 1,500 feet. Lieutenants Meredith, Griffis and Devenney of the 94th were lost. Meredith crash landed in Yugoslavia on the way back and managed to return to the squadron in March 1944. Griffis and Devenney went down in flames south of the target area. Jack Pettus had his problems on this mission. As the squadron was jumped he couldn't jettison his belly tanks. Compounding this he was losing power and had a small fire in his right engine. He headed for the deck, yanking on the tank release handle and feathering the right engine. Fortunately none of the Germans followed him. The tanks finally dropped and, with right engine feathered, he headed back toward the Adriatic strafing a train on the way.

The 71st had similar problems. Major Horace Hanes, the squadron leader, went down with one engine on fire and bailed out successfully. F/O Lee Rowe was hit in the pilot's nacelle and fell out of control. Reed Owens called that he was bailing out but no chute was seen, and Lieutenant Corbett failed to return. During this action the Petdog squadron, all four planes of it, went through a variety of maneuvers to avoid, very accurate flak - accurate enough to jar a P-38 into a vertical position. Separated by flak and cloud layers, they were aware of the enemy attacks but could not locate the other squadrons. They did not come out even in this affair but did account for three FW 190s destroyed, one Me.109 probably destroyed and four 109s damaged.

Major Hanes returned to the group some two and one half months later. He had bailed out seconds before his plane exploded and landed high in a tree. Unable to free himself he hung helpless, frustrated and cold until friendly Yugoslavs freed him by chopping the tree down. He was then sheltered, transferred to the Partisans who moved him gradually, in between fire fights, to a supply field from whence he was flown back to Italy.

Group 'Brass' at Salsola. Top L - R Maj. Sposato (Med.), Col. Robert Richard, Group CO, Lt. Jacobsen 94th C.O., Bottom L - R Capt. Butler 27th C.O. Capt. Wiseman 71st C.0. (Air Force photo)

The next day, January 8th, the air echelon moved to Salsola north of Foggia drastically low in operational aircraft. With all assembled now at Salsola there were just thirty-two operable P-38s in the entire group. Something needed to be done soon or the new Fifteenth Air Force might be faced with another Schweinfurt or Regensberg. The 1st had been very badly outnumbered at Weiner-Neustadt and had a losing hand against the Luftwaffe.

About this time the rest camp on the Isle of Capri began to be used. Pilots were sent in small groups, usually three or four after perhaps twenty-five missions. Enlisted men were sent on a seniority/longevity basis, although this also varied greatly. Capri was an enjoyable place. Located off the tip of the Sorrento peninsula below Naples, it was, and is, one of the most beautiful islands in the Mediterranean. This opinion was shared all the way back to the Roman emperors who summered on the island. Sightseeing was enjoyable, just wandering around the island, boat trips to the Blue Grotto, hotels with clean sheets and good food. For the adventurous there were donkeys to be ridden. The Hotel Quisisana was a popular spot as was the Belle Vista. Rumor had it that there were girls on the island, but for most of us the rumor could not be confirmed. The stay at Capri, or any rest camp, amounted to only a few days but it was a real break from daily missions, C-rations, dehydrated food, washing and shaving out of helmets and four holers. In short, rest camp was exactly that and everyone enjoyed the time off.

1. Vrilakas had relatives living in Athens at this time.
2. From the 155 Autonomous Fighter Group.
3. Remnants of the Italian Air Force continued to fight alongside the Luftwaffe into 1944.

11

FOGGIA

A P-38 Lightning of the 94th Squadron lifting off Foggia airfield past the control tower. (Ken Sumney)

Foggia is accorded separate chapter status because it seemed more like a permanent station in comparison with most if not all of the earlier locations. The 1st Fighter Group was assigned its Salsola base a few miles north of Foggia on the road to San Severo.

Salsola was originally a fairly sizable farm - a Duchy in the olden days. Later it became part of the original Luftwaffe airfield complex on the Foggia plain, the same complex strafed so effectively by P-38 groups in August of 1943. The airfield consisted of a single north-south dirt runway, covered with pierced steel planking. Taxiways and hardstands were also PSP and roadways were simply dirt tracks, which, in rainy weather gave us a quick but lasting impression of Italian mud. Immediately to the west of the landing strip a broad, low hill swept up. Here the three squadrons were located together with group headquarters. The hill was topped by badly damaged stone buildings. The group set up in these buildings including the chapel and a sort of auditorium called the Barn which we used for movies and shows.

January 1944 in Foggia brought forth two imperatives; one was mud the other was cold. The mud quickly got so bad that boards had to be laid where there was much foot traffic. Taxiing an airplane on a muddy pierced steel taxiway could be a little bit like ice skating. If one were not very careful the P-38, with propellers whirling, could slide right into the tail of the one in front with disastrous results.

To ward off the cold we had our 100 octane stoves, one for each four-man tent. These stoves were simple affairs, probably dangerous enough to turn any safety engineer's hair grey, but effective heaters. Each stove consisted of half an oil drum upended over some rocks and/or small pieces of armor plate. The fuel, 100 octane gasoline, was contained in a drum or belly tank outside the tent. The fuel was conducted inside via copper tubing and dripped onto the plate or rocks. A small petcock on the tubing controlled the flow. The proper flow rate was enough to keep a good fire going in the drum. It was chancey, not to say explosive to light, but it was really the only way to keep warm.The stoves could be a hazard though. Danny Darnell's tent burned to the ground one day, and while no one was hurt all the belongings went up in flames including a pair of Danny's boots straight from Texas. In time a version of the stove became Government Issue. Upper and lower iron sections clamped together, sort of a squared off

27th Sq. S-2, L - R Gaines, Krook, Dempsey at Salsola. (Krook)

clam shell. The fuel, the same 100 octane, was dripped into a small bed of sand, where it saturated and vaporized. Lighting was accomplished by tossing in a match.

Shortly after locating at Salsola the medics learned (from the local Italian doctors) that the occupants of Salsola #3 before us, the Luftwaffe, had experienced a severe malaria problem. A swampy area and pond north and northwest of the field was a breeding place for mosquitoes. The area was sprayed with oil once or twice using modified P-38 belly tanks. This helped, but the best solution was to drain the area. This was accomplished using local labor and plows together with some of our own heavy equipment to dredge a drainage canal some three quarters of a mile to the Salsola River.

Foggia was literally surrounded by airfields. The 14th Fighter Group (P-38s) was north of us at Trioli; the 31st (P-51's) was also north at San Severo; the 82nd was southeast of Foggia at Vincenzo and, of course we were ringed by heavy bomb groups. The Fifteenth Air Force rapidly built up to a maximum strength of twenty-one heavy bomb groups (six B-17; fifteen, B-24); seven fighter groups (three P-38, four P-51) and one photo recon group flying F-5s (modified P-38s). All medium bomber units were transferred to the Twelfth Air Force and from this point onward the 1st Fighter Group was destined to escort only B-17 and B-24 aircraft. Shortly after the first of the year General Jimmy Doolittle was transferred to head the Eighth Air Force in England and was replaced by Major General Nathan Twining.[1]

The group had very little time to settle in. There was an escort mission on the 9th, stand down on the 10th and another escort mission on the 11th; this latter to Piraeus, Greece. On this mission the 1st and the 14th escorted four bomb groups and just short of the target area part of the bombers turned back. The 14th turned back to cover the retiring bombers leaving the 1st to go on to the target - all sixteen aircraft. This time the enemy attacked the group with twenty-five to thirty aircraft. We claimed two Me.109s and one probable and all of ours returned safely.

Late in January Captain James Hagenback left the 94th to return to the States. Jim had been squadron commander since September 1943 and had led the squadron through a hectic and kaleidoscopic four months. He had completed sixty-two missions and he was replaced on January 24th by William Jacobsen.

For the remainder of January, weather permitting, the bombers concentrated on enemy airfields and marshalling yards with the P-38s providing escort. On the 27th the airdromes around Marseilles, France were hit. These fields were the origination point for German long-range bombers which harassed Naples and other Italian targets. The same treatment was accorded airfields in the Udine area of northern Italy on the 30th to complete the month.

Bob Vrilakas recalls an unusual January encounter. He was slogging through the rain and mud on his way to the flight line when he noticed a group of attentive people standing in a circle in the pouring rain listening to some fellow standing on a bench talking to them and cracking jokes. He was scarcely recognizable in a pair of muddy fatigues, but it was Joe E. Brown, entertaining as best he could under the circumstances. Most celebrities came over in good weather with considerable fanfare plus supporting cast, but Joe E. Brown had just appeared at the base and was going around entertaining any little group that wanted to listen. It later turned out that he had done this all over the Mediterranean Theater.

Joe E. Brown, beloved American entertainer. Despite the loss of his flyer son in the war, Brown traveled endlessly to visit servicemen. He was in Europe, the Pacific and China. (Karpowicz)

A 94th Ftr. Sq. P-38H-5, Serial No. 42-67091, with long-range auxiliary tank visible, being prepared for the next mission at Foggia. (Ken Sumney)

In February 1944 a group of pilots was dispatched to England to get some more badly needed airplanes. They were transported in a B-17, picked up the Lightnings at Nuthamstead near London and returned via Newquay, Gibraltar, Algiers to Foggia arriving on February 12th. Fifteen pilots made the trip and returned with new P-38H models. The ground crews faced the sudden but welcome job of getting the new ships combat ready. In mid-January the 94th Service Squadron was once again attached to the group as it had been at Mateur. This aided in servicing the new P-38s. In addition to our pilots flying to England to bring back more aircraft we once again gained from the 82nd. The 82nd also brought back used H models from the 55th Fighter Group but, on return, were directed to turn some of the ships over to the 1st.

The mention of ground crews working to bring the new H models up to snuff reminds us that the 1st Fighter Group experience was not all missions, flying and fighting. Maintaining the planes in the air became a battle of wits, scrounging and cannabilizing particularly when we were down to such a pitiful handful.

On this subject, Harley Barlow, a 94th pilot, noted that in all his missions he had never had an early return. His crew chief was Herb Green. Herb's pilot following Keg Barlow was Walter Pregmon. Walter also had zero early returns in fifty missions. This sort of mutual cooperation and admiration was mentioned by several people. Ed LaClare had no returns in sixty-seven missions. His crew chiefs were Alton Long and, later, Bob Taylor. Warren 'Beans' Campbell had one return in sixty-eight missions flying #68. His crew chief for most of those missions was Hollie Baker. Beans says that the single abort was his fault, a burned out starter and fire. The author's solitary return in fifty-three missions aboard Crew Chief Everett Jacobson's P-38 was due to a conked out radio. Bud Miller, who preceded me with Everett, also had only one return in fifty-three missions (coincidentally). And these outstanding records were by no means exceptions.

Early returns were certainly not unusual. When ships were available we flew at least one spare per squadron who flew along for an hour or so to replace anyone who, for whatever reason, had to abort.

Herb Green recalls the daily routine of a crew chief:

> "The day began with a very early wake up call from the flight chief. Arriving at the flight line in pitch darkness we usually didn't know what the mission would be. Escort meant we would need to install drop tanks and fuel them; bombing meant that the ordnance people would bring out the bombs to be hung and, of course, the armament people would be on hand to give the guns one last check. The support people would arrive at the plane after the preflight was complete. Normal procedure for preflighting would be to start with the left engine and look for anything out of the ordinary, paying particular attention to any evidence of oil leaks or coolant leaks. The engines were in-line or liquid cooled so coolant leaks were common and also critical. Inspection continued on down the engine to the wheelwell, looking for leaks, low tires, anything out of the ordinary, continuing down the boom to the coolant radiator. This was a critical area for leaks, also checking on the controllable doors. We checked the tail surfaces for proper operation, continued around the tail section and up the right boom to the right engine.
>
> "With the external inspection complete the next step was to climb up on the wing to check fuel caps, turbos, canopy and windshield. Now we were ready

27th squadron ground crews "sweating out" the mission, Salsola, 1944. (Bob Share)

for engine run-up to check the engines, hydraulic pressures, surface control operation and wing flap operation.

"During the engine run-up we would check for proper mixture control operation, propeller function and fullpower operation. This latter meant running each engine at full power making certain that the ship didn't start to skid-all in near darkness."

During one of Herb's early preflight experiences in total darkness he looked out to his left as the engine reached full power and thought the whole left boom was on fire. He shut everything down and jumped out only to realize that what he had seen was a red hot supercharger.

"Many times all of this was not completed without one or more malfunctions which meant a redoubled effort to get the plane ready for the mission. This might mean changing a set of spark plugs while trying to hold a flashlight and stand on an upended fifty-five gallon drum in a freezing wind. About the time the normal preflighting was finishing the other crew members would be arriving having already been to chow, and would finish up any remaining work while the crew chief went to get some food. Usually by the time the preflight was complete and everyone had been fed the flight chief would come along with the mission information at which time the external tanks would be hung and fueled or bombs hung or a combination of the two. The armament people had hoists to hang the bombs, but it was people power to hang the tanks. Usually two crews would help one another because it was very difficult for just two people to hang the large external tanks. At this point also all the tanks would be topped off. The worst problem would be a change of plans, i.e., replace external tanks with bombs or vice versa. The former was the more difficult because for a long time we had no pumps to defuel the external tanks.

Landing gear change, 94th Sq., Salsola. (Bertoglio)

"Once the mission was airborne it was time to lend a hand to those whose ships were out of commission for whatever reason, engine change, major overhaul etc., or go about the business of washing clothes, working on the tent area or just throwing a baseball around. We were all very concerned for our pilots and, when the flights were returning we were all anxious to see our plane and our pilot return. All the ground crews had a close relationship with their pilots. It was *my* pilot and *my* airplane.

"When the mission returned our day on the ground began again. There would be a repeat of the preflight only in much more detail. Engine cowlings were removed for close scrutiny of the engine, again checking for evidence of oil or coolant leaks or hydraulic leaks. The pilot would fill out the maintenance form as well as the flight report. We would discuss all aspects of the performance of the airplane as well as some aspects of the mission. They wanted us to be informed as much as we could be about the mission."

Despite Herb's reference to the close relationship between the ground crews and the pilots, there had to be some aspects of two different worlds in the relationship. The pilots arrived in the squadron, flew their missions and, if they were fortunate, went back to the States, or, in a few cases, on to group and/or wing. Some pilots were able to complete their tours in a few months. Excepting the originals, very few were in the squadron more than a year. The ground crews on the other hand, were in the squadron for the duration which could mean as long as three years. Add to this the Air Corps separation of ranks by living quarters and areas, and the indignity of mail censoring, and two separate worlds became a reality.

Periodically each officer in each squadron was given a clutch of letters written by the enlisted men, each letter to be scanned for content to be sure no military secrets were divulged. Reading someone else's mail was not a comfortable thing to do. The readers didn't relish it and no doubt the writers didn't either. In any case the Germans seemed to know far more about us than could be gleaned by scanning some guy's letter. In retrospect it was surprising how much the Germans knew about our various organizations and how little many of us knew.

It must have been interesting for a crew chief to deal with what the medics termed a "timid pilot". In a medical summary written in late 1944 the medical staff remarked on the need to, "nurse along weak or timid pilots to complete their tours or be transferred out." Individuals "contaminating to other personnel" (whatever that meant) were rapidly removed. The report goes on to say that there were at all times one to five pilots in that category in the group. The medical report is somewhat terse, but it seems clear that there was the occasional problem pilot who had to be dealt with for whatever reason. Such a pilot would have been a problem for the ground crews - too many early returns - but would have been even more of a problem for those with whom he had to fly. Fortunately these cases were few and fairly obvious.

27th GI's Salsola. Top L - R Weitz, Herrington. Bottom L - R Landry, Krook. (Krook)

In February the weather set in with a vengeance. The group didn't fly at all for the first nine days of the month, however, the down time was put to good use. As Cecil Quesseth, a 94th pilot, recalls from his February 2, 1944 letter home, "We got some lumber a while back and built a good floor and sides for our tent so now we're living in style, so to speak." And again on February 9th from Cecil's letters: "Tonight all the boys in our tent are staying in and reading or writing letters. We have electric lights - this squadron has a small generator - and a gas stove so we are quite comfortable. We also have a table that we had made by an Italian carpenter and a few stools. The wind keeps shaking the tent sides a bit but I believe we have it pretty well staked down."

The above quote sums up everyone's main contact with reality, writing and receiving letters from home. There were the usual card games, ping pong games - when we had a ball to play with - but nothing could beat letters from home be they from family, girl friends, or friends. The latter were usually from someone in another branch of the service in another location and one could compare bitches. But in order to receive letters you had to write, and that too was a comfortable way to pass an evening. In any case, mail call was always a high point of the day.

Continuing weather down time was beginning to get on everyone's nerves. If the weather over the field was good it was bad over the Alps and vice versa. The communal project in progress during January and February was the construction of squadron clubs. In the 94th area the stone building was put up using Italian laborers. A tile floor was installed, some furniture obtained locally, a ping-pong table was added, and quite an attractive bar was designed and built. The clubs were

Club 71, the 71st Sq. Officer's Club at Salsola, Foggia. (McCann)

a much appreciated addition. They provided a good place to go to relax, play ping-pong, bridge, hearts or poker or to have a drink when there was something to drink. The club provided a good and really the only alternative to sitting in one's tent. Club construction went pretty much apace in each of the squadrons.

27th Squadron records also include a foot race and another in their continuing football games, both during the month of February. The foot race involved M/Sgt Charlie Wibirt and T/Sgt Red Everett. The race was based on an argument over which was the oldest. The bet was for $10.00. They were of in a flash running neck and neck when "Moose" Wibirt stumbled and fell. Red finished the race and picked up his winnings. Said Wibirt after the race, "I've been scouting that big redhead for two years and that's the fastest he's ever moved."

In another football game the enlisted men finally beat the officers by the score of 12 - 0, Sergeant Ladson scoring both TD's. The lineup for the officers included: Goebel, Reynolds, Rafael, Lienau, Rodolff, Ferrin, Maloney, Cooley and Sprengel. For the EM's Schneider, Wilson, DiLorenzo, Harvey, Meacham, Carlton, Martensen, Ladson, Pasteris and O'Leary. Those lineups add to nine officers and ten enlisted unless someone was a sub.

On February 10th, at long last, the 94th and 71st flew a dive-bombing and strafing mission to Montero Tondo, Italy. Twelve P-38s dropped twenty-four 500 pound bombs on railroad sheds and tracks then followed up by strafing truck traffic. This was evidently a composite mission, Lieutenant Thomas Collier, a 71st pilot flying a 27th squadron plane, HV-X the oldest plane in the group, was killed when he was caught in the bomb blast of the plane in front of him. Overall six bombs hung up and were finally shaken off.

On February 14th the group provided escort for B-24's bombing Verona. On the 19th the P-38 escort flew for one hour and was recalled. On the 22nd the 1st Group reached their rendezvous point at Klagenfurt, Germany only to find the B-24s were no show.

The last few days of February saw a return to activity with missions flown on the 23rd through the 25th. This period was later known as "Big Week" with efforts of the Eighth and Fifteenth Air Forces coordinated. Unfortunately the weather didn't cooperate for either Air Force. The 23rd was an escort mission providing cover for B-24s returning from Steyr, Germany. For the 1st, at least, it was not particularly eventful. On the 24th fourteen P-38s rendezvoused with B-17s once again at Steyr and escorted them out. Lieutenant Colonel Burton McKenzie, Group Operations Officer was shot down and killed on this mission flying as the 71st Squadron Red leader.[2]

On the 25th the 1st Group escorted B-24's returning from an attack on the Messerschmitt factory at Regensburg. There was heavy opposition by the Luftwaffe but, since the group was covering withdrawal, a good deal of the enemy action had waned. The 27th lost Lieutenant William Parsons in this action and very nearly lost John Price. Apparently one or more enemy aircraft made a high speed pass at the last two Petdog ships and hit both. Lieutenant Price was struck in the face by flying glass when a shell exploded in the cockpit. Nearly blinded by pain and unable to see at all out of one eye he nevertheless collected himself and with the help of Lieutenant Frank Williams flew over 400 miles back to Foggia. Price's instrument panel was broken, the cockpit canopy shattered and the armor plate at the back of his head was broken loose by a bullet. The cockpit was covered with blood. With Williams telling him when to lower landing gear, flaps and adjust speed, Price made a good landing in a rainstorm back at base. Lieutenant Price recovered but lost his left eye.

Lieutenant Emmet Gresham, a 71st pilot, also went down on this mission although not from enemy action and the report of his being among the missing created a considerable flap. Lieutenant Gresham was the son-in-law of Lieutenant General Carl Spaatz. Clyde Gabler recalls being inundated with queries. Lieutenant Gresham, flying Blue #3 reported to Blue leader that he was on single-engine just as the flight was approaching the coast of Yugoslavia en route to the target. Something was seriously wrong with Gresham's P-38; the right engine was smoking and his right wheel seemed to be partially down. On the return flight Gresham's plane was slowly losing altitude with the right engine and supercharger on fire. He side slipped to the left, presumably in an unsuccessful attempt to put out the fire, and then went into a sharp dive. The plane hit and exploded in the center of Prvic Island off the Yugoslav coast. Neither Lieutenants Spitler nor Hoenshell, his two escorts, saw a chute.

As it happens Gresham did manage to parachute out of the burning P-38 and was promptly captured by a pair of unfriendly Yugoslavs, not Partisans. Who they were is not certain, but they probably had some sort of bounty in mind. In any case in mid-afternoon the three of them stopped for a rest, one departed leaving the second one guarding Gresham. Unbelievably, the watchdog fell asleep.

Gresham took his time, looked around and ended up conking the fellow on the head with a rock. He headed south as quickly as he could expecting the worst at any minute. Nothing happened and he was picked up by a Partisan band the next day, whisked to Bari by plane and the back to the squadron, reporting in on March 12th much to the relief of all concerned.

As February came to a close the 1st Fighter Group had managed a total of six escort missions, one dive-bomb and strafe mission behind the lines at Anzio and one air-sea rescue trip around the limitations of winter weather.

The combined efforts of the Eighth and Fifteenth Air Forces were believed to have dealt devastating blows to German armament production during the "Big Week" period. Either we overestimated the damage or underestimated the recovery capability of the Germans; whatever the reason German single-engine fighter production recovered rapidly.

Cajus Bekker in **Luftwaffe War Diaries** gives German single-engine fighter production as shown on the following page in the months immediately following "Big Week":

After tent living in North Africa the 71st Officer's Club was resplendent with fireplace, tables, chairs, hollow bomb casings for ashtrays, a parachute canopy overhead and a/c recognition models hanging above. (McCann)

1944 Month	Me 109	FW 190	Total
February	905	209	1114
March	934	373	1307
April	1011	461	1472
May	1278	482	1760
June	1603	689	2292

Twin-engine fighter production, while in considerably fewer numbers, increased in parallel. The statistics give an idea of the speed with which German industry recovered from what was thought to be extensive bomb damage.

A group construction effort during these early Foggia days deserves description. This was an effort that concerned us all, the chapel. Seeking a suitable location for religious services, Chaplain Morford found the ruins of what had been a large stable measuring some twenty-two by seventy feet. The walls were crumbling, the roof completely gone and animal refuse deep on the floor. After a thorough cleaning and airing, easily done with no roof, reconstruction began. A new roof of belly tank crate lumber was first. Flooring consisted of limestone blocks salvaged from bombed buildings. Lime and sand plus whitewash restored the walls. Seats consisted of metal crates used to protect bomb fins in transit. Some of the other accouterments were ingenious and really quite fine - almost ecumenical if that word is appropriate. Records describe the following:

Pulpit and communion rail—fabricated from wooden crates used originally to contain fragmentation and incendiary bombs. (USA)

Cross—105mm howitzer shell casing, opened, rolled out, cut and polished. (British)

Flower Vases— 88mm antiaircraft cannon shell casings (German)

Candle Holders— 20mm cannon shell casings (USA Aircraft) with 40mm shell casings (British) as bases.

All the metal was solid brass which polished beautifully. When complete the chapel could seat about 200.

With the weather gradually improving, the group flew a total of thirteen missions during March; twelve were escort missions, one was a strafing mission to the Lavariano and Udine airfields. The Udine raid was reminiscent of the Foggia raid of August 1943 and needless to say the pilots were jubilant. The 1st joined with the 14th and 82nd Fighter groups to mount a massive low-level mission which involved some 90 P-38s flying up the Adriatic at altitudes ranging from fifty to 100 feet. The targets were some six enemy airfields in the Udine-Treviso area of northeast Italy plus other targets of opportunity in the area ranging west as far as Venice. Enemy fighter strength on these airfields was estimated at 235 aircraft.

During the attack over fifty aircraft were destroyed on the ground, only one, a Ju. 52 in the air by Lieutenant Raphael, a Petdog pilot. Throughout the raid the pilots kept a lookout for enemy planes but none appeared. Our claims were as follows: three Ju. 88's destroyed, three Ju. 52s destroyed, three Do. 24s destroyed, one He. 111 destroyed, three Ju. 52s probable, one Do. 24 probable, one T/E seaplane probable, three He. 111s damaged, one locomotive destroyed, one radar station damaged, three motorcycles and five trucks destroyed. A number of other targets were strafed including a destroyer anchored in Venice harbor. We lost one man, Harold Lindhurst, a

B-17 buzz job at Salsola. SWEET SUE in foreground. (Lenhart)

Ace Dick Lee with an abandoned German Dornier 217, Salsola. (Lee)

27th pilot. He flew through a high tension line near the airport probably damaging the right wingtip of his plane. He was last seen flying below 500 feet with the right wing tip severed and left engine feathered. There was some confusion in reporting this mission as to which was Udine and which was Treviso but we did hit them both.

Escort missions in March ranged all over the theater from the Anzio beachhead/Cassino area west to France then east to Sofia, Bulgaria. With improving weather the 1st was flying more although "improving" might be too strong a word. The escort missions were flown but in some cases the bombing was doubtful or cancelled entirely. One dive-bombing mission to Anzio was not cancelled and was particularly harrowing. Eight 94th Squadron P-38s, with Bob Vrilakas leading Red flight and Don Kienholz leading White flight were to fly the bomb line west to the coast then up the shore to Anzio. The bomb line was the front where the Alllied and German armies met. Flying toward the front at about 2,000 feet they were about to turn north when the whole sky erupted in flak. The bomb line was not where it was said to be and the Lightnings had unintentionally overflown German lines at a dangerously low altitude. Talk about sitting ducks. Kienholz called to say his controls were locked and he was going into a spin. Just before bailing out he regained control. Rather miraculously, no one suffered any serious damage and, after regrouping, they continued and completed the mission.

On March 11th the 94th put up eleven P-38s, led by Dick Lee escorting B-24s to Toulon, France. Attacked by enemy aircraft over the target area Lieutenants Lathrope and Geyman each downed an Me. 109. Dick Lee tallied two, making him an ace on his 49th mission. On this same mission the 27th put up eight P-38s and the pilots claimed three victories; an Me. 109 by Francis Lawson, another by Armour Miller and an FW 190 by Merle Brown. Clearly, the 1st was still not fielding full squadrons.

On March 23rd Roane Sias returned to the 94th squadron after having been shot down over Naples on August 20, 1943. He escaped imprisonment twice, received the Purple Heart for injuries and, based on work behind the enemy lines together with information brought back, was recommended for the Distinguished Service Cross.

Lt. Donald Kienholz, 94th pilot, February 1944. (Kienholz)

Late in the month we had another spate of bad weather causing aborts of missions on March 23rd and 24th. On the 25th everything was in a "stand down". On the 26th three squadrons took off; two hours later all formations returned because of weather. Finally, on the 28th, an escort mission was completed this time to northern Italy so the Alps and Alpine weather didn't have to be dealt with. On this mission the 27th was bounced separately by some twenty-five enemy fighters about forty miles south of Venice. In a swirling, diving attack our formation was dispersed to the element level. In a twenty-five minute battle Lieutenants James Rodolff and Kenneth Hartwig were shot down. We claimed two destroyed and one probable. In all of this the other two squadrons continued on with the bombers evidently unaware of the fierce fight the Petdogs were in. On March 29th, on an escort mission to Bolzano, the group engaged enemy aircraft in another fifteen to twenty minute dogfight during which Lieutenant Merle Brown, a 27th pilot was shot down by two enemy fighters as he was attacking another enemy aircraft. He was last seen spiralling down with one engine on fire. On an escort mission on the 30th to Sofia, Bulgaria Lieutenants Barlow and Hopkins of the 94th each destroyed an Me. 109 and 71st pilots destroyed two De. 520's in the same encounter.

During March a number of promotions were noted in the records of the 94th squadron. These included Lieutenant Jacobsen, the squadron C.O. to Captain, Lieutenant Andrew Adams, the Intelligence Officer, to Captain, Lieutenant Ralph Kipper, Operations Officer, to Captain.

Lieutenant Colonel Horace Hanes reported for duty once again on March 20th. He had been Group Operations Officer until January 7, 1944 when he was reported missing in action. He had escaped from enemy held territory in Yugoslavia.

Weather conditions continued to improve in April 1944 and the mission count increased as well. The group flew eighteen escort missions and again, a single strafing mission, once again to the Udine area. The mission on April 16th to Brazov, Rumania, was the 1st Fighter group's 1,000th mission of World War II. Thirty-seven enemy aircraft were destroyed, ten probably destroyed and thirty-five damaged during the month while we lost four pilots.

By this time the P-38J-15 models began to appear. The early J models represented a considerable departure from earlier versions. In the cockpit the two handed yoke configuration replaced the interrupted wheel. Gone also was the individual gun charger. Externally the carburetor intercoolers were moved from the wing to the lower front of the engine nacelle, altering considerably the engine profile. Better (flatter) windshield design reduced distortion, and internal gasoline tank capacity was increased. Cockpit heating was said to be improved but this remained to be seen. The new aircraft included one further innovation - no paint. Gone was the usual camouflage color along with about sixty pounds of paint and some wind resistance.The new, bright, aluminum finish Lightnings really stood out in flight intermingled with the old olive drab ships.

As April began Flight Officers Duff, Mimms and Schweikert were appointed Second Lieutenants. Concurrent with his promotion Jerry Schweikert was appointed Assistant Operations Officer in the 94th squadron. These promotions are included only as a means of introducing comments on the Flight Officer rank. The F/O seems to have been something invented by a committee. It was originally intended to be equivalent to the Warrant Officer rank in the Army but designed for the Sergeant Pilots - of which there were several hundred. In the early days of the war the sergeant pilot rank sometimes led to embarrassing situations, at least embarrassing by old Army standards, i.e., a Sergeant Pilot and a Second Lieutenant co-pilot in a B-25 for example, thus the Flight Officer rank. For some reason the F/O rank persisted long after all Sergeant Pilots were commissioned. Throughout the war a certain number of graduating cadets in each class became Flight Officers while the majority of their classmates became Second Lieutenants. Why is anyone's guess or theory. Supposedly the better students received the second lieutenant commission. In practice that didn't seem to be the case. The F/O was sometimes referred to as the "blue pickle" rank because of the characteristic blue line down the center of the gold bar.

During an April 2nd escort mission to Steyr, Germany shepherding B-24 and B-17 bombers, Lieutenant Frank Lathrope had an interesting early return. He had started back because of mechanical difficulties and was jumped by a trio of Me.109's. They attacked him just southwest of Zagreb, Yugoslavia and chased him for some fifty miles. At least one of them closed and got in a good burst before breaking off. No real harm done but Lathrope had a few moments of concern!

Back at the target area the group, attacking by flights, dove into a gaggle of enemy aircraft massing for an attack on the B-24s. In a twenty-five minute encounter seven Me.109s and two FW 190s were destroyed. We lost none. Captain Ralph Thiessen, a 94th pilot accounted for three of the 109s. Enemy tactics were based on diving through the bomber formation then climbing back in pairs intent on singling out individual bombers which they would try to isolate. The 1st Fighter Group intercepted the enemy aircraft as they climbed back to altitude but three Liberators went down in this engagement.

A 94th Squadron P-38H-5, Serial No. 42-67034, makes it back to base for a crash landing. (John Mullins)

The Fifteenth Air Force was built up sharply during April with the 31st and 52nd Fighter Groups transferred from the Twelfth Air Force. Originally Spitfire equipped, both groups adopted the P-51 as soon as possible (at this point mostly B and C models of the P-51 not the later bubble canopy D model).

Ralph Kipper sketched out the duties of the Operations Officer. It amounts to know thy squadron, that is, in response to orders, and at the direction of the Commanding Officer, evaluate the individual pilots and assign flight positions in accordance with each man's experience and expertise. When should a pilot be moved from wingman (#2) to position #4 or "tail end charlie", thence to element leader and so on. Critical decisions for the success of a mission and the smooth operation of the squadron.

Operations coordinates pre-mission briefings and checks to be sure that all squadron activities are synchronized for the mission. That includes Communications, Maintenance, Photography, Intelligence, Supply and Armament. Operations also looks after training, including formation flying, gunnery, aircraft identification, tactics and generally acts as right hand man to the squadron C.O. The Operations Officer or his assistant was usually the guy who woke us up at some inhuman hour of the morning for pre-mission briefing - hazardous duty that.

On 13 April 1944 escorting B-17s to Gyor, Hungary, the squadrons encountered some 40 enemy fighters. Lieutenants Schweikert and Barlow each destroyed an Me. 109.

On April 18th the strafing objective in the Udine could not be reached because of the weather. Instead the group strafed targets of opportunity in Northern Italy. Trucks, locomotives, cars even a steamroller were attacked.

While escorting B-17s to the Venice, Castelfranco area, Cyril Nolen, a 94th pilot, was shot down and became a prisoner of war. Cy was one of the P-40 pilots who joined the 1st in September 1943 in Tunisia. He was downed on his 52nd mission. According to the records, he had six hours transition time in the P-38 before flying his first mission. Nolen was flying spare that day and got more than he bargained for.

> "No one dropped out so I headed back alone and strafed a train in a switching yard and a machine gun that was shooting back. I heard someone on the 94th was coming back also, so made radio contact but couldn't locate him. I accidentally saw a gun emplacement and tried to use combat flaps to lower my nose. I was going too fast and went up about 150 feet. At least eight guns at the front opened up and the BARFLY [Nolan's ship] began to get holes in it. The right engine went dead before I could get back down. I

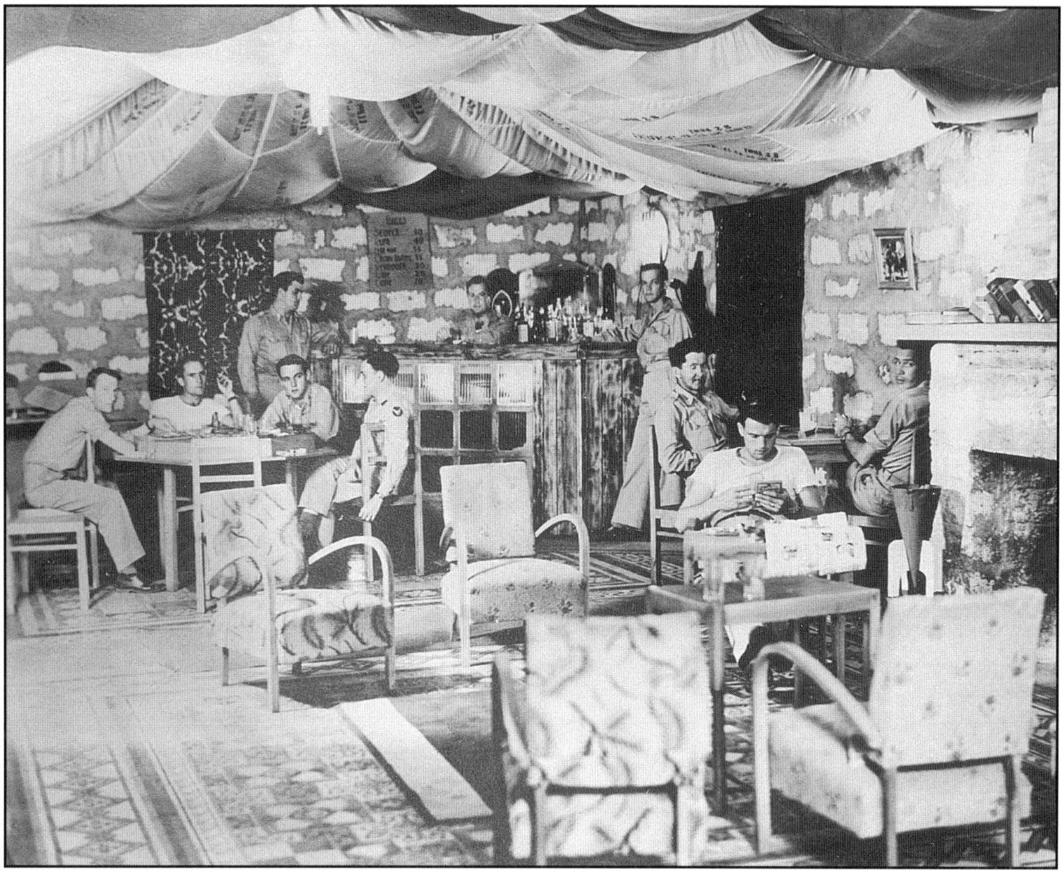

94th Officer's Club, Salsola. Seated at table, L - R Laurent, Hoffman, Dollarhide, Vitale. Standing at bar, Miller, Tattersall. At table, (right) Trost (L) unknown, reading Kirchhofer. (Arrowsmith)

feathered it and then the gas gauges went to zero and the left engine began sputtering and the cockpit was getting smoke in it. I got off a May Day and the BARFLY made it out to the Adriatic and settled in to about forty feet of water, in a bay where the English Eighth Army was facing the German Army.

"There were some Wehrmacht assembled on the shore. I was floating in my Mae West and saw a Spitfire about 8,000 feet looking things over. A German swam out, checked me for a gun and towed me to shore. The guys on shore had me well covered. As soon as we got on shore the English began an artillery barrage. The Germans and I sat behind a stone wall surrounded there being no chance to escape. The next day some P-47s strafed us and we were joined by a P-40 pilot. That night we started north, five German soldiers, two Americans, a German Lieutenant and a driver. Stopping in a small town, Italian partisans warned me about something up ahead. Continuing, the German driver, wearing infra-red goggles and using infra-red lights couldn't see where the Partisans had dug out the road and recovered it. We went rolling down a cliff. Two Germans were killed, the P-40 pilot had his back broken and I was knocked out. I came to on a shoulder of the road with the German Lieutenant holding a broken arm and a Walther (pistol) on me. No chance again.

The Partisans had to stay clear as we were followed by a city bus containing about 40 Wehrmacht. With new guards I got patched up and continued north. Went across the Alps in an electric train with one guard behind me and two on the seat opposite. We arrived in Stuttgart as the English were bombing the place which drove us into a subway type bomb shelter. I was interrogated for about a week. I think they thought I was a P-61 pilot returning from a night mission. The toughest part of the whole business was going without cigarettes."

On April 21st the group flew their first mission against what was to become one of the most difficult and costly targets in Europe-certainly one of the most heavily defended the oil refinery complex at Ploesti, Rumania. It was a deceptively simple exercise that involved neither enemy fighter activity nor losses to the escorting 1st Fighter Group. The Fifteenth Air Force would attack Ploesti repeatedly, yet the refinery kept on functioning after a fashion until Russian ground forces finally overran the area in August of 1944. From April through August the Fifteenth would lose 230 four-engine bombers over Ploesti. The total for fighters was also considerable. The 1st Group alone would lose eighteen missing or killed in action during Ploesti operations. The nine refinery complex was protected by 225 to 250 fighters and 500 to 600 antiaircraft guns of which about half were of the heavy variety, 88, 105 or 128 mm cannon.

On April 23rd the group actually fielded six squadrons, or nearly so, on the same mission. Each supplied two three-flight units. All squadrons provided top cover for B-24s of the 304th Wing to Bad Voslau, Germany. Lieutenant Charles Simpson, 71st Squadron was leading the group on his 50th mission. It was hard to believe that just a scant four months earlier each squadron had been hard pressed to put two flights of four in the air. In fact at Gioia the 94th was experimenting with a formation based on elements rather than flights; that is, breaking the flight in half and covering the bomber stream in two's rather than the standard fours.

All squadrons had their collective hands full at Bad Voslau as each was engaged separately in the target area. For the 94th Barlow and Harris each damaged a single-engine fighter, Charlie Howard shot down a Dornier 215 and Paul Vachon damaged a single-engine fighter. The 71st claimed four Me.109s damaged, one Me. 210 and one FW 190 damaged. The 27th did itself proud destroying three Me.110s two FW 190s and one Me.109. Two FW 190s were probably destroyed and six additional enemy aircraft were damaged. All of our planes returned safely despite a few bullet holes.

By the end of April 1944 the Fifteenth Air Force had reached its ultimate size. This included 21 heavy bomber groups (fifteen flying B-24s, six with B-17s). There were seven fighter groups, three P-38 and four P-51 and one photo recon group. Some of the fighter groups flew P-47s into May but all four groups transitioned into P-51s by early summer. By May 1944 the Fifteenth was operating smoothly and for a change there was a good supply of airplanes. As the weather improved together with the supply of aircraft, flying time increased dramatically. Everett "Bud" Miller flew his first mission on April 16th. On July 26th he flew his 53rd and last. John Allen recorded 23 missions in May alone.

Another sign of the apparent permanence appeared as baseball teams were organized at Salsola. The 94th team was known as the Indians, The nickname harking back to the Indianhead insignia.

The first mission in May turned into yet another weather hodge-podge. The 1st Group took off intending to escort B-17s to the Trento marshalling yards in Northern Italy near Bolzano. They circled the rendezvous point waiting for the B-17s to top the overcast. One group finally emerged and headed for the target, however fighters and bombers again became separated in heavy overcast. The P-38s then overflew a group of B-24s on an entirely different mission to bomb Castel Maggiore. This accidental rendezvous occurred just as some enemy aircraft were setting up to attack the Liberators. After driving off the enemy fighters, the Lightnings, by this time low on fuel, headed for home.

All squadrons provided cover for B-17s and 24s on penetration to targets at Ploesti on May 5th. Additional Ploesti missions followed on May 6th and 7th. On all three of these missions enemy aircraft were observed only at great distances, however the anti-aircraft fire was, if anything becoming more intense, heavy and accurate as the German's increased the protective rings of anti-aircraft cannon around the oil refineries.

On May 10th the group provided escort for B-24s of the 49th Wing withdrawing from Wiener-Neustadt, Germany. The 94th picked up the last B-24 group coming off target and escorted them uneventfully while others shepherded a straggling B-24. At a point near Graz, Yugoslavia ten to twelve enemy aircraft attempted to attack the Liberator. The P-38s responded and a running encounter ensued to a point some fifty miles south of Graz. Frank Lathrope was forced to bail out. Robert Van Sice destroyed an Me.109; Everett Miller probably destroyed a FW 190. Lieutenants Barlow, Hallock and Quesseth each damaged a 109. Lathrope, who returned to the squadron on May 26th, claimed two Me.109s destroyed.

During the Wiener-Neustadt bomb run the flak was intense, accurate, heavy barrage and tracking. One B-24 was seen to explode over the target and one was seen to spin in. No chutes were observed from either ship. Despite the 94th encounters none of the other squadrons engaged any enemy aircraft although most were concerned with shepherding stragglers out of the target area and as close to the Adriatic as possible.

On a May 12th strafing mission in the Po Valley the 27th squadron lost three pilots. Lieutenants Richard Cooley, Robert McIntosh and James Lilly disappeared in heavy clouds in the target area. All were on instruments and evidently lost control in the intense weather. Lilly, reportedly in a spin, bailed out and became a prisoner of war. Cooley crashed and was killed. His grave was found some months later by the American Fifth Army advancing on Bologna. McIntosh was never found but evidently suffered the same fate possibly over the Adriatic.

On May 17th the group was to provide top cover for the 82nd Fighter Group on a mission to strafe enemy airfields at Villa Franca and Ghedi, Italy, then targets of opportunity during the return. Never making contact with the 82nd, the 1st set out to find its own targets of opportunity. They were many and they were costly. The 94th strafed and probably destroyed two Me. 109s and an Mc. 205 on an airdrome southeast of Bologna then probably destroyed two Mc. 205s on an airfield southeast of Ostiglia,Italy. Gun emplacements, railroad stations, boxcars, barracks and power stations were hit and left burning. However, Lieutenants Schaper and Hammerstrom were hit and lost to anti-aircraft fire near the Reno river south of Bologna. Schaper was flying as squadron leader. Lieutenant Duff, also of the 94th apparently the victim of anti-aircraft fire, crashed just north of Lake Valli di Comacchio. There were no chutes observed.

Theoretically one cannot bail out from an altitude of 50 feet, although surprisingly Schaper did survive to become a prisoner of war. After eliciting only name rank and serial number, the German interrogator became impatient and proceeded to tell Schaper that he didn't know much for an officer, then went on to tell him his group, his squadron and the fact that he, Schaper, was flying someone else's plane that day, and he was.

The 27th swept down the Piaconza-Rimini railroad shooting up freight cars and military traffic as well as airdromes along the route. Hangars were hit at Reggio Emilia, a Macchi 202 fighter was set afire and seven others were either destroyed or damaged. Ronald Delaney was hit and parachuted from his burning plane near Pianoro, Italy. Captain Rafael and Lieutenant Joye were also hit but were able to return to base after emergency landings north of Termoli.

The 71st strafed at least three airfields and probably destroyed five single-engine aircraft and three twin-engine aircraft. The squadron went on to damage a dozen single-engine and four twin-engine aircraft and, in the process, lost two pilots. Lieutenant Franklin Pate was hit by flak and crashed in flames near Reggio Emilia Airdrome. Lieutenant Lyle Morehead was hit near Modena. After fighting the flames for some ten miles he attempted to bail out but was apparently unable to. The P-38 flipped over on its back and crashed near Bologna. While these missions in Northern Italy inflicted a great deal of damage it was evident that the enemy ground defenses packed a terrific punch defending their lines of communication.

1. It is interesting to note that after General Doolittle's arrival in England Eighth Fighter Command was eventually allowed to give up "close escort" in favor of a pursue and destroy tactic with great results. The Fifteenth Air Force never varied from the "close escort" concept.

2. McKenzie had recently transferred to the 1st after a tour with the 14th Fighter Group.

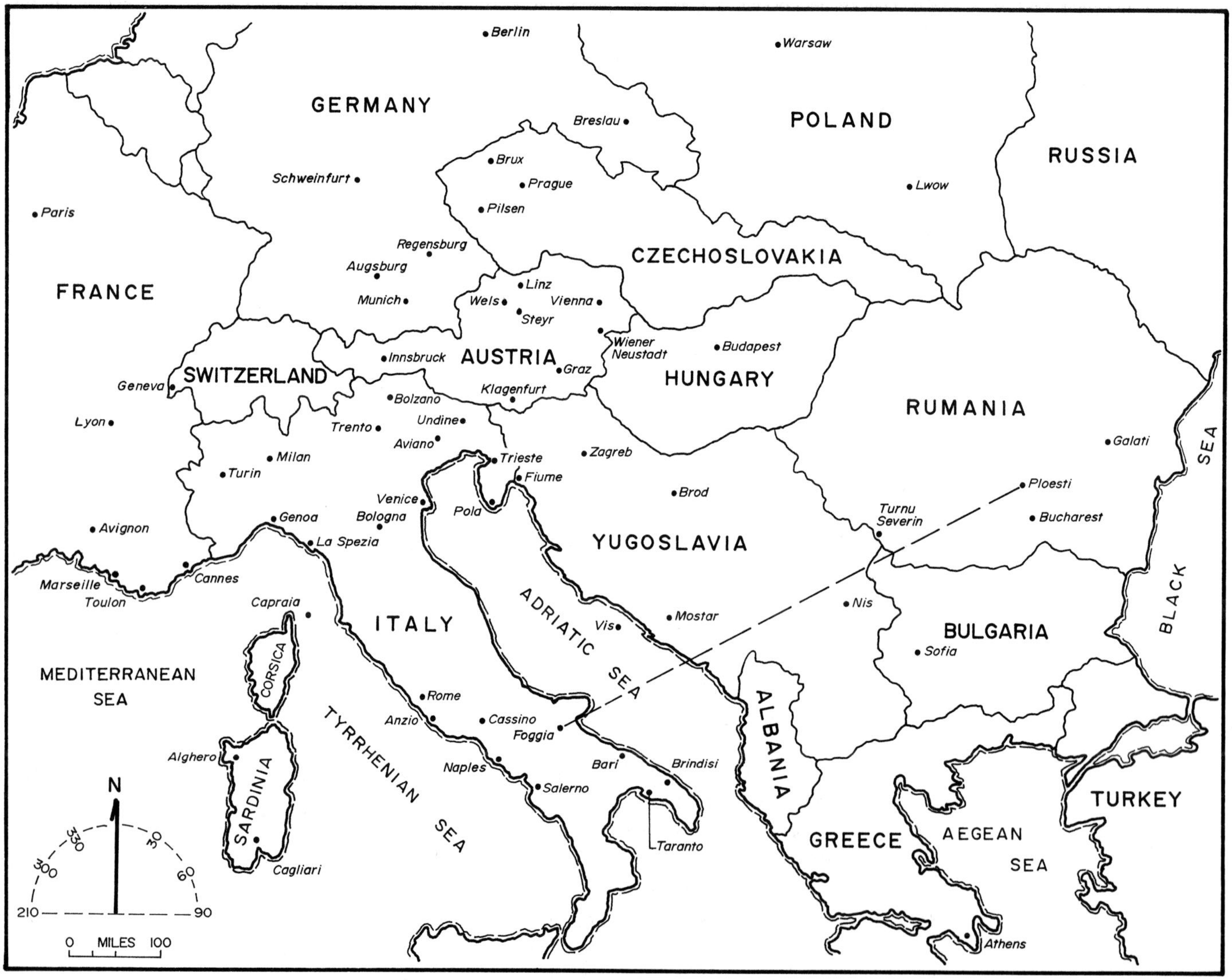

12

DREAD PLOESTI

On May 18th, a mission that was to earn the 1st Fighter Group its third Presidential Unit Citation, the P-38s picked up B-17s coming off the target at Ploesti. This after making their way through abysmal weather that had forced several of the bomb groups to abort. The Forts were besieged by a mixed bag of enemy aircraft and all three squadrons promptly dove into the attackers. After a twenty to twenty-five minute melee the group had accounted for seven FW 190s, three Me.109s and one Macchi 205 destroyed; five probables and five additional single-engine fighters damaged. During this encounter the B-17s continued on course for home excepting one straggler who was beset by a swarm of enemy fighter aircraft. Lieutenant Cecil Quesseth of the 94th led his flight into the attackers and dispersed them, the P-38s shooting down four in the process.

Three P-38s were hit; two returned to base and Frank Williams, a Petdog pilot, was forced to bail out in the Adriatic about 25 miles east of Bari, Italy. Lieutenant Rasmussen circled to be sure of a fix and Williams was picked up promptly by a rescue launch.

Lieutenant Frank Williams provides his view of the long mission:

"We rendezvoused with the bombers and were about four o'clock to the lead group when I called in two enemy fighters making a head-on pass at them. We did a 180 to the right and started a cross weave. As I went under Sprengel's flight I saw two coming down from three o'clock but they weren't in a position to do any damage to us. Continuing the weaving turn to the left I saw two more at nine o'clock going up so I broke into them and pulled up sharply and saw hits on the second one and also hits on the first one from my wingman's guns. The latter spun down to our level and straightened out. Then my wingman called and said he was in trouble. He was directly under us and there was a flight of P-38s at ten o'clock turning towards us, so I told him to fly straight and we would cover him. I started a cross weave with the other flight and was just passing under them when I felt an explosion under my right foot and a sledge hammer hit me in the leg... the bullet, a 20mm HES (high

Lt. Cecil Quesseth, 94th pilot being awarded the Silver Star and the DFC by Maj. Gen. Nathan Twining. (Quesseth)

The diagonal vapor trails above this 461st Bomb Group B-24 are from P-38s on the attack. (USAF)

explosive shell), landed on the floor by my left foot still burning and gas poured into the cockpit. I called and said I was bailing out but the bullet fizzled out without igniting the gas so I turned off all the electrical switches and tried different combinations on the gas selector valves to try to stop the the flow of gas into the cockpit. I found my right main and reserve gas lines were broken and also the crossfeed line. When I had everything as good as I could get it, I called my flight and told them to watch out for themselves but I looked around and they were still with me - excepting #2 man.

"There was a lone bomber pretty close with a flight of P-38s covering it so I took my flight over and started a cross weave 'til we got it back with the rest of the bombers. His right wing was streaming fire but it finally burned out so he seemed to be O.K. I then joined the bombers and flew with them 'til we were clear of the target area. I was now on two engines since my wing tip tanks hadn't been hit but when I ran them out of gas I had to go on single engine because all the gas had drained out of my right engine tanks. My hydraulic system was also out and my nose wheel door was open which created a good drag, so on single engine I could just hold my altitude at 16,000 feet. There were high cumulo nimbus clouds with tops above 30,000 so I had to hold altitude. I saved about fifteen minutes gas in my right wing tank in case I should need two engines to get over some weather. The bombers didn't seem to be steering a good course, although I wasn't sure of my exact position so I left them. After about an hour and a half, I started up my right engine to get over some high clouds and also got a homing and found I was 250 miles from base so I went back on single engine and figured from my gas supply I could just make it.

"My left engine supercharger regulator was out which gave me a lot of trouble. My flight was still with me although I told them to leave several times but I was sure glad they were still around. After flying quite a while longer I got another fix and was still over 200 miles from base. I knew then I would never make it so I used the rest of the gas in my right wing tank to gain more altitude. My oxygen ran out somewhere along the way but I don't know when. I started to plan what I was going to do, and since I had plenty of time I thought it over pretty carefully. I informed Big Fence (the homing station in Italy) and my flight in plenty of time and then concentrated on flying instruments since it was very hazy. About 40 miles from the coast of Yugoslavia a couple of AA (antiaircraft) batteries opened up on us but they weren't very good. My control column was jammed a little and I couldn't get it forward so I couldn't change altitude very rapidly—about 20 miles further we were shot at again.

"Over the Adriatic I held 5,000 feet and when I had ten minutes to go got everything ready. About two minutes before I jumped I sighted the coastline which looked pretty good. I notified Big Fence. When the left engine quit I had everything ready and the ship trimmed to stall. I had a hard time getting out of the cockpit but made it on the third try. I slid headfirst over the trailing edge of the wing, clearing the tail a mile. I started to pull the rip cord but the plane was following me around. I waited a while longer and started tumbling rapidly. I couldn't stop it so I pulled and got quite a jolt when the chute opened. I put the ripcord in my pocket, waved to the boys and started getting ready for the water. It came up faster than I thought so I wasn't quite ready. But my chute blew to one side so I didn't have any trouble there. I had a hard time getting my dinghy out because the CO2 cylinder wouldn't come out of the pack til I cut some of the canvas away. So by the time I got it inflated I was pretty tired and sick from swallowing salt water. I tied the chute to the dinghy for a souvenir since I had the 20 mm shell and the ripcord— might as well save everything. About an hour later a PBY came out and orbited south of me passing pretty close three times but he gave up and went home. After almost three hours a B-25 came over and almost went by before he saw me. He circled and about a half hour later a C-47 brought a launch up. They pulled me out but got my chute tangled in the prop wrecking my best souvenir. A shot of whiskey, a cup of coffee and plenty of cigarettes had me back to normal in a few minutes.

"When I got shot up there were two sitting on our tail. A break was called but my radio was out. I was depending on the flight I was weaving with to cover my tail but I guess they didn't see them. Later I tried to contact the rest of the squadron but my transmission and reception were very poor. Tail-end charlie, Rasmussen, took four 20mm's in his ship before he broke but nothing vital was hit."

As a consequence of this mission the 1st Fighter Group was awarded a unit citation and received the following accolade from the "Big Friends" they were protecting:

HEADQUARTERS 463RD BOMB GROUP
Office of the Operations Officer
APO 520 U.S. Army
23 May 1944
Subject: Escort Performance on Ploesti Raid of 18 May 1944
To: Commanding Officer, 1st Fighter Group, APO 520

1. On behalf of every man in this group, I wish to express our heartfelt appreciation and admiration for the superb work being performed by your group. In providing such excellent fighter cover on many past missions and in particular the last one to Ploesti on 18 May 1944 your organization has done what we consider to be as fine a job of protecting heavy bombers as could ever be hoped for. The reason for citing the last Ploesti raid in particular is that we would have undoubtedly lost many more planes than the six we did lose before your arrival, if your splendid pilots had not arrived to completely route the enemy fighters as they did. These actions by your men were enough to warrant the highest commendation even if they had flown to the target area in good flying weather, but by flying through weather that was adverse enough to turn back all heavy bomb groups except this group they deserve even higher praise.

2. I am sure that you and your entire group would be glad to know how safe and secure we feel when we see your beautiful P-38s flying alongside our planes. Once again I say thanks.
For the Group Commander:
Harry Robb
Major, Air Corps

After nineteen months in the Mediterranean war the enlisted troops were finally enjoying a few amenities. The 94th NCO club (Harry McConnell says the club was really called the EM or enlisted men's club because with very few exceptions, everyone was an NCO) bought a grand piano and string bass. The club now had an orchestra consisting of piano, bass, electric guitar and accordion plus a good PA system for vocalists.

Late in the month the group flew two escort missions to southern France (Grenoble and Avignon) and a dive-bombing raid on an airfield in Yugoslavia. Dive-bombing was usually carried out from an echelon formation by flights starting at an altitude of about 7,500 feet although that could vary considerably. There were no bomb sights. One simply pointed the nose of the plane at the target generally using the gunsight to help aim. The bomb, either 500 or 1000 pound, was released at as low an altitude as feasible, perhaps 2,000 feet. One wanted to hit the target and pull out without; a. hitting the ground or b. flying into one's own (or someone else's) bomb blast. Then there was always the troublesome anti-aircraft fire, although this was not too much of a distraction unless it disrupted your aim. Dive-bombing was somewhere between an inexact science and an art form. Luck was also a requisite. On this particular mission to Yugoslavia our target was hangar installations located at the southern end of the airdrome. Thirteen bombs fell short but were within a radius of 2,000 feet, and four bombs were within 1,000 feet and one bomb hit a storage building. Two bombs were duds. The Lightnings strafed targets of opportunity on the way out, mostly railroad rolling stock.

A weapons carrier load of pilots en route to planes at start of mission, Salsola. (Leen)

On the last day of May on yet another escort mission to Ploesti, two Petdog pilots came up missing. Lieutenant Edward Ulrich became separated following a break. Robert Rasmussen was last seen in the target area with a blown supercharger and left engine smoking after an enemy attack which sent him into a spin. Both reported by radio that they were all right but neither returned to base. Ulrich became a prisoner of war. Much later the fate of Robert Rasmussen, killed in action and buried in Rumania, was learned from the letter of a local official, perhaps a coroner, through the Swiss Red Cross.[1]

Edward Ulrich was more fortunate. He crash landed his burning P-38 in a wheat field where he hid while the German's strafed his ship. Making his way to a road he tried to make friends with two Rumanian peasants using cigarettes and his phonetic language card. The Rumanian's seemed friendly and indicated they would help him meanwhile puffing on his cigarettes. Help him they did by turning him in to a German sentry post a ways down the road. He ended up in a prison compound in Bucharest and was returned with some 1,100 Allied prisoners in August 1944.

Overall May had been a grueling month. Despite the occasional double squadron missions, not that many pilots were available. According to the 27th squadron summaries there was a daily average of only 21 pilots available necessitating daily flying for most. A mission was flown every day the last ten days of the month, all but one of which involved four to five hour bomber escort work.

Nonetheless morale remained high due largely to fairly decent living conditions developed over the past few months plus the recreational facilities provided by the Red Cross and Salvation Army in Foggia. Movies, softball, baseball and volleyball also were popular. With a few exceptions the movies were usually pretty low on that list.

On June 4, 1944 the 1st Fighter Group celebrated its second anniversary overseas with a steak and chicken dinner with all the trimmings. The whole group was entertained by a USO troop. In the afternoon the 94th Indians baseball team took on a team of "officer all stars". The Indians won by the score of 12 - 1. The "officer all stars" are mercifully unidentified although there were some pretty good ball players among the pilots. Jim Fairhurst, for one, aspired to Triple A ball before the war.

As the Normandy invasion was taking place on June 6th, the group flew another escort mission to the Ploesti oil complex. Despite several saturation raids by the heavy bomb groups Ploesti was proving to be a mighty tough nut to crack. Although damaged, the oil refineries were turning out substantial quantities of everything from gasoline to heavy engine oil. In addition to very effective smoke screens the Germans had built entire dummy refineries to mislead the attacking bombers. Once over the target area attackers were subjected to extremely heavy antiaircraft protection and an effective if diminishing number of enemy fighters.

On June 10th the Fifteenth Air Force decreed a surprise low level attack aimed specifically at the Romano-Americano refinery at Ploesti. This mission was unique in that the 1st suffered its highest casualties of any mission in WW II.

Lieutenant Armour Miller describes the mission planning:

> "In late May a plan was developed to undertake a low level surprise dive bombing attack on the power station and cracking plant of the Romano Americano refinery, which had thus far escaped major damage from any of the high altitude raids. The plan theorized that such an attack might be accomplished before the German defenses were alerted to fully deploy the smoke screen. The 82nd Group was selected to dive-bomb and the 1st Fighter Group to fly escort to protect them from premature attack be enemy interceptors. The raid's success was dependent on the following:
>
> 1. Complete surprise without detection by enemy radar or radio.
> 2. An extra large disposable fuel tank (a 300 gallon ferrying tank was used) along with a 1,000 pound bomb
> 3. Effective protection from enemy fighter attack to preclude the premature jettisoning of external tanks and bombs.
> 4. Accurate low level, dead reckoning navigation to achieve the target and the designated climb point.
>
> "As selected formation leader I attended an advanced briefing session held at wing headquarters before the mission was undertaken. Our group C.O., Operations Officer, myself and a like complement from the 82nd along with ranking briefing officials of the wing attended. Route, target, formation, defenses and possible losses were discussed in detail. We were not allowed to take maps or write notes and were advised that those would be withheld until the mission morning briefing. We were ordered not to discuss this briefing with any personnel at our group. A schedule date would be decided later subject to favorable weather forecast and additional intelligence information. Certainly the fear that enemy intelligence might get advance notice of this mission was of foremost concern to wing officers."

Forty-eight P-38s of the 82nd Fighter Group took off at dawn on 10 June 1944 from Vincenzo air strip, Foggia with Lieutenant Colonel William Litton, 96th Squadron, leading. Each plane carried one 1,000 pound bomb and one belly tank. The 82nd was escorted by all three squadrons of the 1st Fighter Group as the two came together over Manfredonia at 1,500 feet. The 27th was 1st Group lead squadron with Armour Miller flying

Seeing the mission off, Salsola. (Bertoglio)

as squadron and group leader. The 71st, headed by Lieutenant John Shepard, was middle squadron flying to the left of the 27th. The 94th with Lieutenant Harley Barlow as leader was designated as final squadron.

Both groups headed across the Adriatic, then Yugoslavia, flying as low as they dared given the mountainous terrain. En route twelve 82nd aircraft and nine 1st Group P-38s made early returns leaving thirty-six "bombers" and thirty-nine as escort. Following the Danube they would pivot northeast around Bucharest, climb to bombing altitude and (the 82nd) dive on Ploesti. That, at least, was the plan. The 1st began its turn north at the IP just east of Giurgiu on the Danube and the 27th reached its final checkpoint southeast of Ploesti, more than 600 miles outbound within five minutes of the ETA.[2] The 82nd did not turn immediately so the 1st corrected by turning back to the east. As they did this the 82nd began its turn north. The 1st turned again, however, 71st Blue flight was cut off by the 82nd and had no choice except to continue north separated from the other three flights of the 71st.

At that point the 71st spotted at least six Dornier 217 bombers flying below them toward the north and dove on the enemy aircraft downing six (two by Shepard, two by Hoenshell, one each by Hisey and Flack). However, the three flights were hit almost immediately by at least twenty Rumanian IAR 80s, a single (radial) engine fighter easily mistaken for a Focke Wulf 190. [3]

The 71st could not have been at a greater disadvantage. The enemy had position, speed, height and numbers. Within minutes the 71st lost nine planes and pilots. Willie Flack was killed outright and Lieutenants William Potter, Joseph Jackson, Vernon Baker, Albert Smith, Rudolph Janci and Ralph Hisey went down. Lieutenant Carl Hoenshell shot down at least three of the enemy but was hit by an Me.109 on the way home. Janci and Hisey survived to become prisoners of war. Last to be lost was Lieutenant George Johnson who was probably wounded before he went down.

At nearly the same time, still southeast of Bucharest, the 94th flew through a low flying covey of aircraft. These included four Do. 217s, one FW 190 and many trainer type biplanes. The 94th literally burst through this flock destroying two of the Dorniers and five of the trainers, then continued on to the target area located east of the city of Ploesti.

Both initial encounters involving the 71st and the 94th Squadrons took place just as the 82nd Group, escorted by the 27th, and 71st Blue flight, began climbing to commence the dive-bomb run. Surprise was far from complete but the dive-bombing did start before the German smoke screens could develop completely. The 82nd pressed home its attack on the Romano-Americano refinery and seven P-38s were shot down over the target by a hail of anti-aircraft fire and two more were lost in strafing attacks on the return to Italy. The 82nd dropped thirty-six bombs. A cracking plant was hit plus a huge storage tank and three of the refinery units were damaged in varying degrees. Coming off the target the 1st hunted for and found targets of opportunity. The 27th, strafing west of Ploesti enroute home, tangled with thirty to forty Me.109s, shot down four, damaged four more and claimed two probables. On the debit side the 27th lost four. Captain August Bischoff caught a wingtip strafing and was killed; Harry Noone became a prisoner and James Joye was killed in action while William McClellan was wounded and ended up in Partisan hands in Yugoslavia and returned to Italy late in June. The 94th lost one, Ray Geyman, who crash landed, probably hit by a Dornier tail gunner. He became a prisoner of war. Overall the 1st scored 18 confirmed victories but lost fourteen of its own. Herb Hatch, one of the few surviving 71st pilots, shot down five enemy aircraft this day. Hatch and Hoenshell were both awarded the Distinguished Service Cross, the only two awarded to the 1st Fighter Group throughout the course of the war. Hoenshell nearly made it back. Herb "Stub" Hatch describes the mission from his vantage point of Cragmore Green #3, the element leader of the fourth flight:

"We were briefed that morning very early. We got up around 0400, had some breakfast and went down to group headquarters for the briefing. When we walked in and sat down it was apparent that something unusual was in the air because all the group brass were in attendance. When they went to the map and drew the line to Ploesti all of us kind of went, 'Uh,oh.' And then, when they told us what the mission was, there was absolute silence and utter disbelief on the part of all of us who were going to fly over 600 miles to surprise the Germans in order to dive-bomb the Romano-Americano refinery.

"In the course of the briefing it came out that the 82nd were the ones to do the bombing. We were selected for fighter escort. I can't adequately describe the sense of relief that went through that gathering when we found out that we weren't going to be the ones carrying a 1,000 pound bomb on one side of the airplane and a belly tank on the other - or that we'd be the one's to try and dive into that unbelievable flak.

"Take-off at 0505 went as scheduled. We rendezvoused with the 82nd and headed for the coast of Yugoslavia. Anyone who has flown formation at low level knows the difficulty in keeping a squadron of sixteen aircraft together, let alone three squadrons. Nonetheless, we hit our IP right on schedule south of Bucharest. At that point we began our turn north, dropped our belly tanks and were supposed to begin our climb to altitude to cover the 82nd. As we completed our turn however, we flew right over an enemy airfield and in the airfield pattern were four or five Dornier 217 twin-engine aircraft. Our squadron leader, First Lieutenant John Shepard turned in and went after them and three flights followed him. (Blue flight was, by this time, cut off) The Dorniers didn't last long. I only wasted some ammunition by firing at one of them at the tail end of the little fight.

"At this point we were only 250 to 300 feet off the ground. As we pulled up slightly to turn back north again somebody hollered, 'Cragmore break left for Chrissake!' I looked to my left and there was a whole flock of FW 190s headed in from ten o'clock high. [Actually Rumanian IAR 80s]

"Our entire squadron broke to the left. As I continued around in my sharp left turn a lone 190 came out of nowhere and pulled right across in front of me. He was so close - fifty to seventy-five yards away - that all I could see in my ring sight was the belly of his fuselage and the wing roots. I opened fire with all four .50 caliber machine guns and the 20 mm cannon, and I just damn near blew him in half. That saved my neck because when I rolled out to shoot at that 190 I looked to my right and here comes another bunch of 190s from my two o'clock.

"There were four 190s in the lead. I did the only thing I could do. I turned sharply to my right, pulled up and opened fire again. The leader was 150 to 250 yards away, nearly head-on and slightly to my left. I set the lead 190 on fire with a burst that went through the engine, the left side of the cockpit and the wing root. The 190 rolled to its right and passd me on my left. I didn't see him crash but my gun

A 27th Sq. P-38 takeoff, Salsola. (Bertoglio)

A Rumanian IAR 80C, unit unknown, 1944. On June 10, 1944 they were mistaken for Focke-Wulf 190s. (Dr. Jan Koniarek)

camera film showed the fire and my wingman, Lieutenant Joe Morrison, confirmed that he crashed. Unfortunately the other three 190s in that flight went right over my head and down on the tails of Green flight leader and his wingman. Both were shot down.

"As I continued to turn around to my right, my wingman stayed with me and I saw another 190 right up behind one of my tentmates, Joe Jackson flying as Cragmore White #4. I closed in on that one from about his five o'clock and tried to shoot his canopy off from about 100 yards, but I was too late to save Joe. By then the 190 had set Jackson's plane on fire. Joe's plane rolled over and went in and he was killed. I finally did get my burst into the cockpit area and the 190 followed Joe right into the ground.

"I was still turning to the right, going quite slowly by then, because I had my combat flaps down. I turned maybe another ninety degrees to my right and saw one of our '38s coming head-on with a 190 on his tail. We were still around 300 feet and the P-38 passed over me by fifty to seventy-five feet. I pulled my nose up and opened fire on the trailing 190 from a distance of about 150 to 200 yards. He kept coming head-on and I shot off the bottom half of his engine. He nosed down, still shooting at me and I had to dump the yoke hard to miss him. He was burning when he went over me by not more than three feet, and part of his right wing knocked about three inches off the top of my left rudder.

"As this 190 went over my head, I saw three more making a pass at me from my left. I turned so fast I lost Joe Morrison. I missed my shot that time but when these three had gone over me (they went after Morrison) I saw three 190s diving at another P-38. I snap-shot at the leader from about a 90 degree deflection. I hit his left wing and shredded the aileron. He fell off on his wing and went in. He was so low that there was no chance for him to recover. I kept on going around to my left and shot at the second one which was going away from me on my left. I hit him, but I'm not sure if he went in. I know I knocked a bunch of pieces off his cowling and fuselage but I didn't have time to see what what happened to him.

"I looked up at two o'clock and here comes another 190 right at me. It was too late for me to turn. I just shut my eyes and hunched down in the cockpit. I thought I'd bought the farm right there. But he missed me, he never even hit my ship. I think he missed me because I was going so slow. He overestimated my speed and was overleading me. I started to turn his way and, when he went behind me I continued on around. There was another one out there so I closed in on him. I took aim, fired but my guns only fired about ten rounds and quit. I was out of ammo. I damaged him a bit but he flew away.

"I cannot overemphasize what a melee that was. There were at least twelve P-38s in that little area, all of them at very low altitiude. Somewhere between twenty-five and thirty 190s were also in there. None of us was at more than 200 or 300 feet and some were quite a bit lower. The topography was kind of a little hollow with some hills on each side. It was by far the wildest melee I saw in the sixty odd combat missions I flew. There were aircraft going in all over the place. I heard one guy who had been wounded pretty badly, scream until he went in. It was a wild, wild few minutes. And a few minutes is all it was. According to the mission report from our debriefing the whole

Herbert 'Stub' Hatch, 71st Squadron pilot, who was awarded the D.S.C. for his actions near Ploesti on 10 June 1944. (Hatch)

fight took something like three to six minutes. I had no inkling of elapsed time while it was going on. I was too damned busy trying to stay alive.

"When I woke up to the fact that I was out of ammunition, 600 miles into enemy territory and all alone, I broke out of the area and went looking for some company. In only a few minutes I found one of the other planes in my squadron headed in my general direction. I called the pilot, Carl Hoenshell on the radio and we joined up. About that time I heard my wingman, Joe Morrison, hollering for some help. He was on single engine, pretty badly shot up and would someone please come and help him. So Hoenshell and I turned back to look for Joe. We finally found him down around 200 feet. After we got him headed in our direction we started to climb out of there to the west.

"Joe's airplane looked like a lace doily. The two 190's I had not had time to turn into had gone over the top of me and down onto Joe's tail because he had broken right when I had broken to the left. Joe's ship was flying but just barely. Hoenshell and I were both out of ammunition. The three of us tried to make ourselves as small as possible and headed west. Four or five minutes later another P-38 joined with us. This was Lieutenant John Allen, a 94th pilot. We hoped he had some ammunition. When we called to ask we found that his radio was out and we couldn't talk to him.

"Another twenty-five or thirty miles west, just as we were gaining some altitude we ran into a bunch of flak. Unfortunately Morrison became separated from us again because he couldn't maneuver as quickly as we could to get out of the flak, so we had to turn around and go back and get him again. We nursed Joe along for a long, long time. Finally we got out of Rumania and into Yugoslavia and had climbed to about 12,000 feet. We were S-ing back and forth over Joe beause he couldn't fly as fast as we could on his single engine. As I was turning from one of our Ses I spotted six Me.109s about eight o'clock. I hollered to Hoenshell, 'Bogies high at eight o'clock! He saw them too and cautioned, 'Hold it, Hold it. Joe, hit the deck.' Joe didn't lose any time. He stuck his nose down and headed for the ground.

"Carl, Allen and I held the turn as best we could and, when the 109s broke formation and came at us from six o'clock we turned into them hoping to scare them off by looking like we were ready for a fight, but they didnt scare worth a damn. (This with no ammunition). When Hoenshell, who was leading, hollered on the radio, 'Hit the deck Hatch!' I didn't waste any time doing just that. I rolled my airplane over on its back and split Sed out of there.

"One 109 was chasing me and a couple of others were going after Hoenshell, but I don't know where the others went. There was an undercast beneath us and I didn't have the faintest idea where the hell the mountains were - Yugoslavia is full of them - but there was no choice at that point. The Me.109 was chasing me and I had nothing left to fight with so I went through that undercast so fast I didn't even see it. I was hitting close to 600 miles per hour when I came through the bottom into a valley between two high ridges. The Lord was sure with me that day.

"I kept going. When I was sure I'd lost the 109 I pulled back up over the overcast and started looking around for Hoenshell, Allen or Morrison, anybody. I heard Joe hollering for help but my fuel level was getting down to the point where I couldn't afford any longer to turn around and go back. I continued on toward Foggia.

"When I landed back at home base I was the first member of our squadron to return from the mission. It was noon and my elapsed time was six hours and fifty-five minutes. I don't think I had enough gas to go around again if I hadn't been able to land on my first approach. There was quite a welcoming committee at the revetment when I parked the aircraft. Shortly after I landed Cragmore

Blue flight came in-all four of them. They hadn't been in the fight at all.

"Much later that evening, long after debriefing and after we seven survivors had imbibed a bit of the medicinal alcohol that the flight surgeon was kind enough to put out, who should come wandering into our officers club but Joe Morrison, my wingman. He'd gotten that lace doily across the Adriatic but had to dump it on the field at Bari. He'd hitched a ride on a truck and walked back into the squadron that evening.

"After all the smoke and dust had settled I was credited with five confirmed victories, one probable and one damaged. It's my opinion that the probable was a pretty sick puppy. I have hunch that he never flew again, but I can't prove it."

The Ploesti mission was number 27 for Hatch and his first as element leader. Prior to this he had been flying either number two or tail-end-charlie.

Carl Hoenshell went down in Bulgaria and died of injuries the next day. John Allen made it back to base. Armour Miller assembled elements of the 27th and shepherded them back toward Italy. He took two of the most damaged to Bari, landed, then continued on to Salsola. Elapsed time to Bari, seven hours thirty minutes. Total mission time eight hours, nine minutes.

The 82nd bore the brunt of the always deadly Ploesti flak. The 1st particularly the 71st Squadron was caught at low altitude, at cruising speed by a large enemy force which had every advantage. In a few minutes the 71st was reduced to little more than a flight. Reviewing the chronology; when the 71st was jumped the 27th was climbing to 12, 000 feet east of Ploesti. The 94th had sailed through its covey of enemy aircraft and was climbing to join the 27th. Neither the 27th nor the 94th seemed aware of the 71st fate even at debriefing, noting only that the squadrons became separated in the target area. There was never a suggestion of turning back to assist the 71st because that would have meant abandoning the 82nd pilots. After leaving the Danube and heading toward Ploesti both groups had encountered quite a parade of enemy aircraft. The 82nd had been instructed not to engage any enemy aircraft en route but in a few cases simply couldn't avoid them. 96th Squadron pilots downed an Me. 210, a Hinkel 111 and two single-engine fighters. Two other aircraft were also shot down by 82nd pilots, none of whom jettisoned their bombs or belly tanks in the process.

Despite the valiant efforts of the 1st and 82nd Fighter Groups, the results of the "experimental" June 10th low-level raid on Ploesti look meager when compared to the cost. The 1st Fighter Group lost fourteen P-38s and most of the pilots while the 82nd lost nine: a loss ratio of thirty percent.

The Rumanian version of the June 10 mission is somewhat different. As related by one Dan Vizanty, evidently one of the leaders, they thought the target was their own airfield near Bucharest. They seemed unaware of the entire 82nd Group and the real target, a Ploesti refinery. With at least three languages involved some of the exaggerated statistics can be overlooked, but Vizanty was right about one thing. They caught the 71st by surprise and cut it to pieces. Vizanty indicated that the normal practice of his unit, once an alert was sounded, was to take-off and climb to about 30,000 feet by which time they would be positioned some sixty-three miles north of Bucharest. On this morning they had taken off and reached about 4,600 feet when they were alerted and, "saw the Lightnings approaching at low altitude from the east with twin bombs." [Presumably the twin bombs were belly tanks.] Vizanty goes on to say, "It was our luck to shoot down the commander of the unit [which they did not do]. Our attack was so fast that not one of the hundred American planes managed to fire a shot on our planes still on the ground." An interesting defenders point of view. Vizanty's 100 American Lightnings were, in reality, three flights of 71st Squadron ships, twelve P-38s. As a defender he missed the entire 82nd Fighter Group and two squadrons of the 1st and, of course, his airfield never was a target so it's not surprising that no one shot at it.

Needless to say, the entire group was stunned by the losses incurred during the Ploesti raid. The 71st was hard pressed to field a sixteen ship squadron for the continuing escort missions. The Fifteenth Air Force evidently concluded that dive bombing attacks on heavily defended targets such as Ploesti were too costly which was certainly true. It's surprising that such a mass flight of P-38s was not detected and attacked sooner. Such a mission was not repeated, but the escort missions continued on nearly a daily basis as the weather continued to improve.

With so many missions to Ploesti it's inevitable that at least one would include a bit of macabre humor and also reprise a long-ago similar experience. James Morehead tells the story:

"A funny but not so funny incident occurred on my European assignment. Our G-2 officer wanted to go on a mission. He kept bugging us to take him on a mission. Well, I thought this a rather admirable desire for a staff officer to want to learn first hand a part of his business. So, we rigged up an extra oxygen source piggy-back in my P-38. We took out the armor plate and made him a seat and warned him of the long, cold ride. Oh, no problem. Well, I'm leading the Hat-in-the-Ring Squadron toward Ploesti, pointing the nose toward thirty-five angels and about every 5,000 feet I call back over my shoulder to see how he's doing. OK, OK! Then we come into some enemy activity, I've forgotten whether it was fighter or ack-ack and we are wrapping around with some degree of preoccupation and all of a sudden I feel something on my shoulder. Then I remember Captain Carton, the G-2. I yell, 'Are you OK Cart?' No answer. I turn quickly to see this purple-blue face lying on my shoulder. Holy Moses, at 30,000 feet he has only a couple of minutes to live. So I pull out of all the gymnastics and head straight for the ground. That happened to be another of my mistakes. By holding such a steep angle too long I entered compressibility, that dread condition in a P-38 where the aircraft goes out of control and tucks under in a fateful dive for the earth.

"With old Cart behind me, unconscious, I can't bail out and leave the poor guy so I start all known procedures for getting a P-38 out of compression. I cut and saw the throttles, I kick alternate rudder; I push and pull on the stick but to no avail. I'm heading straight vertical. I try the trim tab and thought I felt a change. Not much but ever so little. I crank back and forth and she starts coming out. Beads of cold sweat are popping out of my pores. The mountains in the area of Sarajevo, Yugoslavia are coming up fast and appear to be my final resting place along with 1st Fighter Group G-2. We shot down to about 8,000 feet at around 500 miles per hour while I, in desperation, milked that P-38 out of compressibility.

"I called Blue flight leader to take over the squadron and continue the mission as I had a patient to return behind the lines. A moment or two later Cart came to life, took the weight off my shoulder and asked feebly how we're doing. It was some time before I told him how close he had come to buying a Yugoslav farm (Air Corps lingo for death)."

1."On the 31st May during an aerial fight that took place in the neighborhood of the town of Pascani and Caciulati, an American twin-engine aircraft was shot down and fell 600 meters from the village of Caciulati. (the area is approximately 200 miles north of Ploesti). At the time the aircraft was falling smoke was coming from it, and we supposed that the pilot was gravely wounded and that he attempted to land but crashed into the ground after losing control of the aircraft. Prior to the crash the aircraft was seen to regain control for a short time then falling into the ground at an extraordinary speed (800 to 900 km per hour). The remains of the aircraft were scattered within an area of about 220 yards. The aircraft was buried to a depth of between fifteen and 20 feet and burned for two days. Muffled explosions were heard.

"On the 9th of June we learned that the remains of the aviator were not yet buried and the representative of the French Legation and of the French Institute told Mr. Cristesco, Rumanian Professor at the French College of Bucharest, to verify these facts and to contact the Rumanian authorities of the village with a view to digging out the aircraft and the remains of the pilot.

"The 9th and 10th several crews of soldiers and laborers were employed digging this ground which had become as hard as a brick due to the fact of the fire from the aircraft.

"On the first day we recovered a great quantity of metal and the following objects: five deteriorated machine guns, one of which was the 20 mm cannon, a great quantity of ammunition, two propellers, two radiators, three-fourths of the engine, armour plate of three-quarter of an inch thick, the burnt radio receivers, and a small quantity of white melted metal.

"On the first day most of the work was done at the area where the engines were buried fifteen to twenty feet in the ground. The following morning, on carrying out further work, we discovered a small metal plate indicating the type of aircraft.

"We also found a small medical kit. Between the two engines, the cockpit was discovered. Near this area was also found the parachute pack partly burned. At a greater depth we found a pocketbook containing a calendar, bank notes (American lire issue for Italy), an immunization card, two ration cards which were three-fourths burned. From these ration cards the aviator was identified as 2nd Lieutenant Robert P. Rasmussen, 27th Fighter Squadron.

"The following day, at eight o'clock in the morning, a coffin containing the remains of Lieutenant Rasmussen was taken to the cemetery near Pascani. The Reverend Father Gouillard, Assumptionist, Chaplain of the French Colony of Caciulati, recited the prayers and blessed the grave. The group was composed of the following: Lieutenant Jean Correge, Mr Robert Disse, Professor of the French Institute and two local Rumanian Army Cadets.

"Lieutenant Rasmussen did not receive military honors from the Rumanians as shortly before the burial, an alert was given. At the time of the burial, a heavy formation of American aircraft was passing over the cemetery going from north to south. This coincidence permitted Lieutenant Rasmussen to receive the farewell of his comrades in arms."

2. Per Armour Miller's notes.

3. A Polish language monograph on the IAR 80 by Konarski and Pecko has information from the Rumanian sources regarding the engagement. Rumanian Air Force Grupul 6 Vinatoare was stationed at Popesti-Leordeni Air Drome near Bucharest.

"The Lightnings which took off from bases near Salsola and Triolo were soon detected by radar. Twenty-six IAR 80C fighters belonging to squadrons number 58, 61 and 62 were at standby. Warned of the approaching enemy, the Rumanian pilots took off within seconds against the Lightnings...The Romanians lost three pilots."

13

THE NOT SO SOFT UNDERBELLY OF EUROPE

Major James Morehead brought an interesting perspective to the 1st Fighter Group and to this history. Amazingly enough he came full circle in terms of assignments. In the Southwest Pacific at the start of the war he was a member of the 17th Fighter Squadron.[1] The 17th was one of the original squadrons of the 1st Fighter Group until it was detached and sent to the Philippines in 1941 and replaced by the newly created 71st squadron. Major Morehead left the balmy Pacific area and arrived in Italy in a driving, icy rain to join the 71st squadron in 1944. He traveled rather rapidly from 'tail-end charlie' orientation missions to squadron operations officer to group operations officer. He comments on the differences in enemy traits.

"Since I fought the Japanese, the Germans, the Koreans and Chinese, people ask about the differences and there certainly are differences. The Japanese taught their pilots that once they latch onto an enemy their attention must not deviate until the enemy is destroyed. I think that that was actually good policy for bomber pilots and I believe that is the best way to be effective for them. But they (the Japanese) used to make their fighter pilots sit in a barrel for hours in order to impress them on boresighting your target. I considered this dumb and it helped our greenhorn kids blow them out of the sky. For the Americans and the Germans almost the reverse was true. To me the Italians, Rumanians, Koreans and Chinese were all rather unpredictable. I felt all except a few Germans and Americans were too careless with their lives. Nearly all of them would fly several moments in a combat zone straight and level without clearing their tails."

Warren 'Beans' Campbell recalls an incident that occurred on a June bomber escort mission to Munich one of those, "now why did I do that?" things that people just do. Beans was flying Blue #3 and Blue #4 had aborted leaving Beans, in effect, tail-end charlie.

"Nearing the target area we ran into fifteen to twenty Me.109s. They were above us so the squadron leader started a slow climb to meet them. It was about this time that Blue #2 started to drop behind apparently with some kind of engine trouble (Blue #2 was flying a P-38G while I was flying a J model) Blue #2 dropped further and further back and I throttled back to stay with him and protect him.

Lt. Warren 'Beans' Campbell 94th pilot, Salsola. (Campbell)

Lt. Tom Maloney is telling his crew chief, Sgt. Richard Abbott, about his fifth aerial victory, an Me. 109 near Ploesti, on May 31, 1944. (Maloney)

"Suddenly he seemed to have corrected his problem and with his new-found power, started to pull back into position. While his G model was older it was also lighter and therefore, faster in initial acceleration. As a result as he pulled back into formation I was left all by myself with enemy aircraft bearing down on us.

"It was at that moment that I looked back at about seven or eight o'clock high and saw an Me.109 diving down on me, machine guns blazing. I could see the tracers falling behind me but they appeared to be getting closer. It was here that I displayed my exemplary bravery..... I ducked and put my head between my legs. (I didn't think at the time that if I were hit that would just provide a longer body target) Anyway, after what seemed like an eternity, probably about three seconds, and nothing happened, I dared look up and to my surprise and horror, all I could see for a fraction of a second was the right wing and part of the underbelly and tail surfaces of the 109 not more than 100 feet in front of me, probably even less.

"At first all I could think of was that if I had my guns firing as he passed in front of me I would have blasted him out of the sky. Then later as we were returning home after the battle, I began to think of how close we had come to a mid-air collision...........at our speeds, his probably about 300 and mine in a climb about 210, only a fraction of a second. It was several hours after we landed that the cold chills stopped running up and down my spine."

On June 15th the 1st Fighter Group had another of the costly strafing raids. This time to Orange Plan de Dieu Airdrome in France. The entire formation of three squadrons flew on the deck over the sea, rising only to cross the mountainous regions of Corsica. At the target the 71st flew cover at 10,000 feet while the 27th and 94th strafed the airdrome. Lieutenant Robert Spitler, a Cragmore pilot, shot down one FW 190 which was stalking the strafing squadrons. Lieutenant Don Greenley of the 94th damaged an Me. 109 making a head-on pass at the strafing P-38s. On the ground four Ju. 88s, one FW 190 and two Me.109s were destroyed. As the strafers came off the airfield they sought targets of opportunity and attacked rail traffic. Lieutenants Dugglesby and Crandall, both 94th pilots, went down in flames on the mission, hit over the airdrome by anti-aircraft fire. Strafing enemy airfields was nearly always a hazardous venture and no respecter of experience. The fields were literally ringed with heavy machine guns and 20mm and 40 mm rapid fire cannon. These could put up a hail of crossing fire. Some of the German tracer shells looked for all the world like orange golf balls as they came up.

The very next day, an escort mission to the Labau oil blending plant near Vienna demonstrated that the high level escort missions were not without their own brand of hazard. This mission was particularly costly for the 94th. About thirty enemy fighters were encountered on the way to the target. In a twenty minute running battle Phillip Smith destroyed one Me.109, Bob Van Sice destroyed one and damaged another as did Charles Dollarhide. Chester Heien accounted for one destroyed and one probable, both 109s and Jim Fairhurst damaged an Me.109. However, in the fight we lost George Loughmiller, Thomas Vitale and Samuel Barnes, all shot down in the Lake Balaton area. Loughmiller was a former B-17 pilot who had transitioned to fighters. This was his first mission and he did not survive it.

On this date the German jet fighter, the Messerschmitt 262 made its initial appearance without any engagement. The first sight of that plane always

Royal Nyby, 94th pilot brushing teeth, Salsola. (Arrowsmith)

27th Squadron P-38 close up view. (Campbell)

provoked disbelief - nothing could go that fast at that height - but there it was, it could and it did. On the return trip Royal Nyby and Jim Fairhurst stayed with a straggler B-17 with two engines shot out. The B-17 headed for Vis but was losing altitude and wouldn't make it. Over the Adriatic the B-17 crew began to bail out. Nyby and Fairhurst moved in to get the identification numbers on the tail of the bomber then count the chutes. All important information for mission debriefing.

Jim Fairhurst recalls for us his own experiences on the Labau mission:

"Our squadron was flying high altitude escort for the heavies, B-17s and B-24s. Many heavy bomber boys can tell you about droves of 'Jerry' fighters that we met over Lake Balaton that day. On the morning of June 16 Captain LaClare, Operations Officer, came into our tent and said, 'Barnes, Fairhurst lets go, briefing at 0615.' Barnes and I were real good buddies; we had gone home together before coming overseas. 'Dusty' was his nickname. Dusty and I left our sacks and started to dress. As we were pulling our flying suits on and checking our .45's he turned to me and said, 'Say Jim, if I ever go down take care of my things and whatever fits is yours.' I said, 'OK Dusty and the same goes for me too.' There was nothing unusual about the conversation, most every 'Joe' in combat makes that pact with a favorite buddy.

"We took off at 0745 that morning, Dusty flying Green #2, myself flying White #2. Our squadron was made up of four flights, Red, White, Blue and Green with four ships in each flight. The flight leader #1, his wingman #2, the element leader #3 and his wingman #4, also known as 'tail end charlie. Our group hit Lake Balaton at 28,000 feet, four minutes ahead of schedule. As we swung into a 360 degree turn to the left, calls started coming in on the radio. 'Bogies nine o'clock high, Bogies two o'clock high.' Then, 'Springcap squadron-drop belly tanks, 'Break right!' The fight started.

"My first impression was a flurry of slender black arrows slanting through our formation. We were in a scrap with the yellow nose Me.109s, the Yo-Yo boys. They had a reputation and were known as Yo-Yo boys because anytime anyone tangled with them they seemed to be going straight up or straight down. Just after our first break I looked off my left wing in time to see three P-38s get it. One burst into flame all over and around the cockpit (Loughmiller), the second went into a slow spin to the right trailing coolant (Barnes); the third had stopped flying and appeared to go by me as if some giant hand had hurled it across the sky. I saw the pilot of the third ship hit the silk OK. For a minute or so it seemed as if I had been watching all this from a seat in a movie house back in the States. Finally I got with it and raised a little hell of my own. On the way home 'Ques' our squadron leader (Cecil Quesseth) called for a check in: Green 2 (Barnes) failed to check in along with Red 4 and Blue 2 (George Loughmiller and Thomas Vitale) When I heard this I was sure praying that 'Dusty' was the one that hit the silk OK but he was not."

On June 24th Lieutenant Marlyn Kobiske, a 94th pilot, was killed in a crash just a mile or two north of the Salsola base. He had just taken off for a practice training flight. Kobiske was a replacement who had arrived at the group directly from advanced training with no operational training whatsoever, and consequently had little or no flying time in fighters of any type. It befell the more experienced pilots in the group to try to train these new fellows when they could. This was difficult for all concerned in the busy summer months when flying missions was practically an everyday affair. Kobiske's accident was more understandable; it was his first time in a P-38, however these local accidents were not infrequent, usually fatal and usually inexplicable. Such accidents usually occurred with inexperienced pilots. Whether some malfunction caused a loss of control or one simply became disoriented will never be known. Overall in the 1st Fighter Group nearly one loss in ten was a local accident of some sort involving a training, orientation or testing flight.

As the Italian spring progressed to summer the base at Salsola took on more and more the aspects of a small town. One of the farm outbuildings had been converted, much like the chapel, into a small auditorium called the Barn. Movies were shown in the Barn once or twice a week, some pretty good, some pretty awful. Usually after the movie we would watch some gun camera films taken from the past few missions. Some were good, some not so good depending on the amount of vibration, sun position, etc.

Those of us assigned as new pilots discovered that Italy in the summer was; a. very hot, b. very dusty, and that tents for four with cots for beds were not really what one could call posh accommodations. We also discovered, based on observation, that a lot could be done with belly tank crate lumber. We began with a wash rack to contain the usual helmet wash basin, some flooring and some shelving. This later progressed to some fairly elaborate cabinetry on an amateur basis.

"Sgt." Tony Antonio Castriota, 27th Squadron mascot, Salsola, summer 1944. (Bob Share)

Living in the Italian countryside presented its contradictions and contrasts. At least until early 1944 the Italians were our recent enemies. In actuality most were anything but. They were friendly, desperately poor and invariably seemed to have relatives living in Boston or San Francisco or Chicago. So they would say with much arm waving and an occasional, 'finito Benito,' accompanied by a shrug of the shoulders. A visit or two to Foggia convinced us that we were indeed strangers in a strange land - but not so strange as Africa. We were exposed to strange customs, strange attire, a strange language, bombed out buildings, hordes of child beggars, etc. Malaria was endemic in the area which made a daily Atabrine tablet a requirement and turned each of us a rather strange yellow color. The single Italian we had daily contact with, at least in the 94th, was a former prisoner of war named Gino who was employed as a waiter/handyman in our mess dining room. Gino's command of English was somewhat limited. When someone complained about the palatability of powdered eggs or dehydrated potatoes, his standard reply was a shrug of the shoulders and the remark, "Spik son of a bitch sergeant!" That seemed to cover any situation and relieved Gino of any further responsibility. The 27th also had their Italian contact, a young man about twelve years of age whom the 27th 'adopted'. Antonio Castriota, called 'Tony', was outfitted with a cut down sergeant's uniform and cap and looked quite the young soldier.

Being 'on location' for such a length of time gave rise to a great deal of ingenuity particularly on the part of the enlisted ranks since they were on site seemingly forever. Bob Naismith, one of the original 71st pilots, passing Salsola in a P-51, stopped in for a visit with Bucky Harris, C.O. of the 94th, and two of his ex-crew chiefs, 'Cajun' Johnson and 'Mouse' Fredrici. Fredrici in particular, seems to have been a genius at fabrication. The following is Naismith's description of Fredrici's living quarters:

> "He had designed a structure around the configuration of engine packing boxes. He had rigged a P-38 window crank beside his bunk and attached it by cable to a hinged window on the opposite side of his room so that he could control ventilation without getting out of the sack. He had positioned two drop tanks on scaffolding at roof level on the outside. One held aviation fuel with a line leading to an oil drum furnace with a heater vent leading inside and with a petcock controllable from his bunk. The other tank was for fresh water. One line from it went through a coil inside his furnace then through the side of his house; the other went directly in. Fredrici had set up a wash

stand against one wall. Central was his basin, a landing light reflector from a P-38 with a drain hose connected to the lamp insert hole. Feeding into the basin were two more petcocks, one for hot water, one for cold."

In the 94th Stanley Karpowicz, a mechanic and amateur photographer, devised a hand agitated washing machine which bore an amazing resemblance to a hand operated 1890's style washer stored in the author's grandmothers barn in Massachusetts. Hers was wooden mostly, Stanley's was metal, mostly. Stan also documented much of the 94th history with literally hundreds of photos taken with his trusty Leica.

During July 1944 the major effort was in long range bomber escort. We flew 23 escort missions all of them deep into enemy territory including targets in France, Northern Italy, Germany, Austria, Rumania, and Hungary. Additionally we flew two long range fighter sweeps. Of the latter, one took us to the Ploesti-Bucharest area and the second to Northern Italy to the area of the Lavariano Airdrome in the Po Valley. Lieutenant Howard Micheli, a new 94th pilot, failed to return from the Lavariano sweep, his first mission. After leaving the target area some of the squadron was making for the island of Vis with at least two damaged aircraft including that of Lieutenant Paul Vachon, the squadron leader that day. Micheli was with the squadron till well out over the Adriatic and then he was not. He may have been low on fuel and set out for Italy on his own. If so, no one saw him leave. So far as anyone knew he had not been hit. He simply was no longer there and no one had any idea what had happened to him. In addition to Micheli, three pilots from the 27th squadron went down on the same Lavariano fighter sweep. Lieutenants George Eldred and Leo Lemons were seen to crash land in the target area. F/O Thomas Wood was also reported to have crashed. Lavariano airfield was heavily strafed on the raid particularly the hangar areas and the surrounding anti-aircraft emplacements.

Major James Morehead, a Cragmore pilot recalls experiences with strafing raids at Ploesti.

"We were trying to reduce the air defenses over Ploesti by strafing aircraft supply sheds on a heavily defended airfield that did not lend itself to a low approach. As I led the sixteen planes into a dive from a couple of thousand feet the ground and sky lit up with fire, smoke and myriad streams of tracers. Someone on my right was hit, someone on the left blew up. I found a good target and gave it the full four .50's and 20mm fire. Suddenly my left engine belched smoke and oil drenched my left empennage. I held the trigger down until I swept low over the target, then I chopped throttle and scrambled to cut switches and feather the engine to try to avoid the coming fire. With full throttle on one engine the plane yawed dangerously, but maybe this was good for as far as I could tell I was taking no hits. Quickly cranking trim, I could tell she was straightening up some and I was slowly acquiring a balance between power and trim and she was becoming a flying machine again. With a careful turn I got headed back toward Italy instead of Russia. The 600 miles home was alone and with more thrills yet to come.

"Somewhere in Eastern Yugoslavia I came across a Panzer column of Germans riding on top of their tanks, lolly-gagging along. When I opened up with my remaining ammunition there was a wild scramble. I don't think I did much damage except perhaps to their dignity. I made it back to Italy that day and got her on the ground in one piece, but I was alone in our five man tent that night, a testimony to the German efficiency in the defenses of the most important target of WW II. The 1st Fighter Group caught it hard in the air battles over Ploesti in 1944. Twice in my tour of duty there I spent lonely nights. Twice the other four pilots in our five man tent were lost. I think I was about to get a complex."

'House on the hill' haven of Mouse Frederici, 71st Sq. crew chief. The magazine is *Better Homes and Gardens.* (Hansen)

Further recalling the hazards of low level attack in the Po Valley, Major Morehead continues:

"It started off pleasantly enough, in fact kinda comical from my viewpoint. We had strafed a row of trucks along a road on the way to the target. As I peppered the column one truck was screeching to a halt and out the door came flying a big, fat German soldier tumbling head over heals into the bar ditch, the truck still rolling down the road. Further on we came upon a group of buildings, and as we popped over the structures, I was

Gun control. The business end of the P-38 displays some of its four .50 cal. machine guns and one 20 mm cannon, Salsola, 1944. (Karpowicz)

expecting to find a worthy target. Sure enough a dozen men dashed from one of the buildings. I had nosed down expecting a target and they came right into my sights. I was pressing the trigger when I recognized a red garment, then a woman and a child. Oh me! I had come very close to slaughtering a group of Italian civilians.

"Arriving over the German airfield at 20 feet the small arms fire was like a hail storm in reverse with white streaks and fire balls going skyward searching us out. I hugged the ground like a sidewinder. I might have claimed more Germans with the propellers than bullets. Unfortunately an inexperienced pilot was flying directly in front of me but much too high above the ground.

"On missions such as these we practically take an oath of secrecy—nobody utters a word on the radio. In this case I considered breaking silence and telling the boy to fly lower for up there where he was, he was offering a target. I didn't. I wish I had. A moment later he was struck in the right wing root and flame showed through. The boy was as if paralyzed. He flew straight and level and I followed him closely. We were now coming up on the mouth of the Po River. My casualty was now in a gentle climb in the zone of great danger. Various defense batteries were pecking away at us as the fire in his right gas tank continued to burn. Now, over Venice, all the port guns had a crack at us but we flew through it and out over the Adriatic. The flames burned deeper, but we were away from enemy terrain. I thumbed the mike to yell, 'Bail out!' but I was too late. I had felt it in my bones that he was playing on luck too long. At that very instant the right wing broke loose. Also, apparently the pilot had gotten hold of himself and was making plans to bail out on his own. As the beautiful twin-tailed fighter rotated around itself in its last throes a parachute canopy appeared amidst the tangle. Then the whole mass struck the water. Hoping beyond hope, I started to circle the wreckage. Then I got the message from an ever widening cloud of red coloring staining the blue of the Mediterranean Sea."

Vis, the island for which some headed after the Lavariano raid, was a very useful refuge. A small tract of land (some thirty-three square miles in area), it's one of the outermost islands of the Dalmatian archipelago well out in the Adriatic from mainland Yugoslavia.Taken in 1944 by Yugoslav partisans, it became a focal point for joint Allied operations in the Balkans. It also became a haven for numbers of disabled American bombers and fighters damaged in the long penetration raids or low on fuel and unable to make it back across the Adriatic to Italy.

On July 16th on an escort mission to Wiener-Neudorf, Austria, Lieutenant Jack Hughes was lost apparently as a result of a mid-air collision with a P-38 of the 82nd Fighter Group. Hughes plane was seen descending with both booms severed and the right engine on fire. On the same mission Lieutenant Charles Morris crash landed his P-38 at Salsola and emerged uninjured.

One of those real tests of endurance occurred during this operation. Lieutenant Frank Holcomb of the 27th squadron was hit by flak well into Austria. He was severely injured; one arm was badly broken and he was alone 400 miles from home base. Everyone within ear shot mentally willed him to make it as he flew toward Italy. He was in shock, bleeding, and judging by the way his voice faded in and out, getting weaker yet he did make it back, landing in Bari near the 26th General Hospital.

Probably the major mission of July occurred on the 18th. On an escort mission to Friedricshafen, Germany, B-24s of the 49th Wing were scheduled to bomb the Manzell Dornier Works. Through some change in plans the B-24s were not at the rendezvous point, but B-17s appeared. Approximately ninety enemy fighters attacked the bombers from six o'clock slightly low. The P-38s flying behind and above the bombers dove into the enemy fighters. The German pilots, predominantly in FW 190's were intent on the bombers and persisted in their attacks. The 27th squadron, flying on the left of the group formation, led the assault and destroyed fourteen enemy aircraft and damaged or probably destroyed eleven others. We lost none although there was some damage to the Lightnings. The enemy aircraft approached from the two oclock position to attack the bombers from the rear. The 94th flying as group lead was out of position. The 71st, flying to the right of the 27th was cut off by six o'clock attacks to which they responded. Thus the 27th took on the main German thrust and gave more than a good account of themselves.

Lieutenant Bill Caughlin, a Petdog flight leader, in describing the encounter said, "I shot at least eight of them and two of them went down. All the Jerries that I didn't get my flight got. Every man in my flight got at least one enemy." With the 71st boxed out and the 94th too far in front, the 27th was badly outnumbered but had the advantage of speed and altitude. Still, consider the odds: fourteen Petdog pilots attacked at least seventy-five enemy aircraft and beat them off. Not to say the B-17s came out of it unscathed. The 27th pilots reported eleven B-17s shot down. A large number of chutes were reported. Also reported were several instances of enemy aircraft firing at the parachuting air crew. This mission and others at about this time reinforced a German tactic not new but now more consistent. The enemy fighters were extremely persistent in attacking the bombers but avoided contact with P-38s and P-51s as much as possible.

The term yo-yo boys was never more meaningful. The Me. 109s and FW 190s would split-S away from the American fighters but rebound immediately in their yo-yo fashion to again attack the heavies particularly any straggler bombers or bomb squadrons. It seems we rarely saw an enemy fighter that wasn't going straight up or straight down. And of course going straight down in a P-38 was something done only with considerable risk, a fact the Germans knew just as well as we did. Unfortunately this situation was not remedied until the appearance of dive brakes in the late model P-38Js.

Lt. Bill Caughlin, July 1944, 48th mission, 2 victories. (Caughlin)

Crew chief, Hollie Baker (right) and assistant at Salsola, summer 1944. (Campbell)

One example of a German fighter not zooming up and down was experienced by author John Mullins on a mission in July. This was reminiscent of "Beans" Campbell's ducking experience:

> "On one of my early missions the squadron leader had called a break left after the usual drop belly tank call. I had switched to internal tanks and, just as the break was called, one engine quit. Reaching down to be sure I hadn't just cut off the gas in switching tanks, I managed to knock off my oxygen mask. Fumbling with all of this I suddenly realized that those were tracer bullets under my wing. I rolled over and split-S'ed into a layer of clouds beneath us. When I came back up, now with both engines operating, everyone had disappeared. There were no enemy fighters, there were no P-38s, I couldn't even see the bombers. After circling a while and seeing no one, I finally set course for home feeling not a little sheepish. Landed, still feeling sheepish, and got debriefed feeling even more sheepish, in fact, downright embarrassed."

The huge bomber escort raids of 1944 had, in some ways, a terrible beauty about them from a fighter pilot's point of view. Arriving at our rendezvous point we would circle anxiously looking for the bomber formations and suddenly there they were, a few thousand feet below us in a column extending as far as the eye could see. Hundreds of four-engine bombers, each squadron and group in its own box formation boring ahead toward the target, sometimes underlined by contrails sometimes not. And suddenly the flak started, ugly black explosive puffs with an occasional red flash as something was hit. At that point your glad your in a fighter and not sitting there plowing through those deadly black explosions.

Lt. Col. Francis 'Bucky' Harris, 94th C.O. (Leen)

94th pilots Walter Pregmon (left) and John Mullins, Salsola. (Raner)

July 28th proved to be the exception to the rule- as usual. Escorting B-17s and 24s to Ploesti the 27th squadron was jumped by some 40 enemy fighters. Coming out of the sun they ignored the heavies and went round and round with the Petdogs for a few hectic minutes.

Major Francis 'Bucky' Harris had assumed command of the 94th Squadron in mid-July. Bucky Harris was one of the originals and as a member of the 71st Fighter Squadron he had flown the Atlantic as a part of the BOLERO Mission. Completing his missions in North Africa he had returned to the USA, then rejoined the group in May, 1944, thence to the 94th in July relieving Captain Edward LaClare who had been acting C.O. About the same time Major Francis Pope assumed duties as Commanding Officer of the 27th Squadron relieving Captain Thomas Rafael.

In July the 1st Fighter Group was advised of the fate of Lieutenant William Grieshaber and Cyril Nolen. Italian partisans reported that a member of the Italian Carabineiri (a type of Italian police) saw Grieshaber's plane crash land in the ocean about 100 yards offshore during the mission of August 10, 1943. Grieshaber swam ashore, was captured by German soldiers who searched him, took his possessions and then shot him in cold blood. At about this same time we received confirmation that Cyril Nolen, shot down on April 20th, had been captured and was now a prisoner of war.

During July the rotation program made definite progress. Four officers and thirty-seven enlisted men, all veterans of over twenty-six months overseas duty, were transferred to replacement centers for reassignment to the USA. That still left the group with ten officers and 581 enlisted men who had seen continuous duty overseas for over two years.

As an alternate prize to Stateside furlough, Rome had been added to Capri as an official rest camp location. The Air Corps had taken over several hotels in the Eternal City, most of which were entirely unscathed. Rome was also in considerable contrast to Capri as a rest camp. The city was large, cosmopolitan and a mecca for sightseeing of all kinds. We trooped to the Vatican, took in the ancient Roman ruins and ate and ate and ate. The hotels featured excellent food and even such niceties as strolling strings.

Social aspects of camp life at Salsola also began to improve. We had three stage shows featuring French and Italian entertainers on an outdoor stage built beside the Barn. Lily Pons appeared at the stadium in Foggia and the Irving Berlin production of **This Is The Army** was staged at a theater in Foggia. Tickets to these affairs were sort of by lot but a good number of our people attended. Irving Berlin visited the 1st on July 7th, had lunch with Colonel Richard and went for a P-38 piggyback ride with Howard Standridge a 71st pilot. Mr. Berlin said he enjoyed his ride which included a couple of slow rolls.

Despite improvements, living conditions at Salsola in the summer of 1944 were nothing to write home about. The days were very hot and very dry. When the tent sides were rolled up to catch the breeze the dust also came along. Then too, water drawn from a Lister bag standing in the sun all day didn't do much for thirst. It was hot enough to shave with. On the other hand the weather was great for flying. We rarely had a stand down during the height of summer. One additional

71st Pilot, Howard Standridge (left)and passenger Irving Berlin, renowned composer. (Karpowicz)

Lt. Stan Lau, 27th Ftr. Sq. scored three aerial victories in July battles bringing his final score to 4-0-2. (Caughlin)

break in the summer was swimming in the Gulf of Manfredonia. The coast was an easy drive and we had a fairly nice beach rather handy.

In the latter part of the summer the P-38J-25 model began to appear at Salsola. This modification of the J series represented the greatest change yet at least from the pilot's point of view. The two major differences were the addition of hydraulic aileron boosters and the placement of dive brakes under each wing. The aileron boost made an amazing difference. It was similar to what we now call power steering on automobiles and improved the control sensitivity of the P-38 enormously. During first take-offs in the J-25s the ship literally fluttered as it left the ground due to the pilot overcontrolling despite his best efforts not to do so.

The dive brakes were electrically operated, wedge shaped affairs (when extended) located under each wing. With these brakes the P-38 could, at long last, be put into a near vertical dive without encountering compressibility. In the past the controls could very well freeze in high speed dives until the plane got into thicker air at lower altitudes. In truth most dives were not really vertical although they seemed to be. Tony LeVier, Lockheed's famed test pilot, once estimated that the reported vertical dives were probably more like sixty or seventy degrees than ninety. No matter, at high altitude the P-38 could get into a very high speed dive very quickly. One in which controls surfaces did not respond, and in that interval pilots could and did panic. Why it took till the summer of 1944 to make such a critical modification is puzzling. The dive brake modification was designed and actually approved early in 1943. By the time the brakes were installed in the J models over half of all P-38s had been constructed. So goes bureaucracy, wartime notwithstanding.

We should inject here the experience of Lieutenant Warren 'Beans' Campbell who took a P-38 up to 27,000 feet on a test hop:

"I got bored, curious to see what would happen if I rolled it on its back and took my hands and feet off controls. I found out! Didn't get past straight down. High speed stall... wouldn't pull out, wouldn't roll. Last time I looked the indicated air speed was 575 mph. Finally, I got some feel in the elevator trim tab and pulled out over the Adriatic at about 4000 feet."

After flying 68 combat missions 'Beans' went on to fly the 'Droopsnoot' P-38 for MAAF. The Droopsnoot was a P-38 modified by removing the guns installing a Plexiglas nose and either camera or bombsight plus a bombardier (a small one). In this way a whole squadron of P-38s equipped could bomb while flying straight and level toggling their ordnance with the bombardier flying in the nose of the lead the Droopsnoot.

On August 1st the 1st Fighter Group had a full scale military review; so far as is known the only one of record during World War II. Officers and men in Class A uniform passed in review then stood at attention as three Presidential Citations were awarded along with several individual decorations. Brigadier General Dean Strother, Commander of the XVth Fighter Command performed the honors together with Colonel Richard, C.O. of the 1st. And then it was noon time and back to work and to muse about how we all got through a full military review. For some of us the military commands and responses to commands were reasonably recent, but for others it had been years since the parade jargon had been heard. Nonetheless everyone agreed we made a smart looking outfit and it was a stirring sight to see everyone standing under the flag with the P-38s in squadron colors lined up as a backdrop.

August 1944 proved to be a momentous month for the 1st and the war effort in general. The month began as July had ended, a series of hot, dry, dusty days ideal for flying, and the flying continued to be the long-range escort missions guarding heavy bombers on their deep penetration raids into Fortress Europa. The exception occurred on August 6th when we flew escort for the 14th Group as the latter dive-bombed the airfields at Plan de Dieu and Valence France. In contrast with the disastrous June mission escorting the 82nd to Ploesti this one was both uneventful and successful.

Returning from a Ploesti mission on August 10th the 94th came upon two Re. 2001s setting up for an attack on a straggler B-17 near Nis, Yugoslavia. The attackers broke off as the P-38s turned into them. Dick Arrowsmith chased one down to the deck and damaged it at the very least.

1. Serving in Java with the 17th and later in Australia and New Guinea with the 8th Fighter Squadron, Morehead scored seven victories flying P-40s.

Its P-38 nose heavily modified, the DROOPSNOOT is shown with Warren 'Beans' Campbell pilot, upper left and Charles 'Chuck' Morris, 94th pilot checking for fit. Sgt. unidentified. (Karpowicz)

14

OPERATION ANVIL

Air echelon detachments began their move to Aghione airfield, Corsica on August 10, 1944 in anticipation of the invasion of Southern France, code named ANVIL. The airfield at Aghione was a dusty, dirt strip on the East Coast of Corsica. Pilots were housed in an abandoned two story hotel a couple of miles from the airstrip and ground crews were in tents close by the field. Chester Heien remembers no doors and no window glass. The kitchen set up, at least for the pilots, was open air just outside the hotel. Eating was pretty much an al fresco affair. Bathing more often than not, was in a small stream which ran behind the hotel. As the author recalls, his first meal in Corsica was some sort of jelly sandwich and a can of warm beer. For invasion support both the 1st and 14th Groups moved to Corsica while the 82nd stayed in Italy for bomber escort duties.

Herb Green's recollection of Corsica operations reads almost like a Norman Rockwell painting:

> "Missions were a dawn to dusk operation on Corsica so the ground crews were on duty on the flight line pretty much all day. A creek ran close by the runway complete with trees and a rope swing tied to a limb. All this thanks to the people who had occupied the field before us. In any case it was a cool relief to swing out over the creek on the rope and drop into the water. A lovely break between engine overhauls, refueling and rearming.
>
> "Prior to leaving Salsola for Corsica the pilots had 'escape' photos taken. Each posed in what was assumed to be a costume and pose typical of a native of Southern France. Assuming that photos would be hard to come by should we be shot down and forced to take to the French countryside, we would provide our own. 'Beans' Campbell's photo is the only one that has surfaced after all this time. The author recalls that he posed in an old civilian jacket probably borrowed from the local Italians. I had a cigarette dangling from the corner of my mouth because, 'that's the way the French smoke'. I was also told to feign deafness."

27th pilots living quarters, Corsica. Two pilots on left are unidentified. From center (L - R) Lts. Bart Jones, Tony Noonan, Galen Rhoades. Critical items on table include toilet paper, mess kits, beer and candle. Rhoades was killed in a mid-air collision at the end of the Corsica operation. (Air Force)

Bath time at the old swimming hole on Corsica. (Karpowicz)

'Beans' Campbell recalled an incident on one of the early patrols from Corsica, this on August 15th:

"I led a flight of four on an armed recce over Southern France where we dive-bombed a bridge and then patrolled an assigned area off the southern coast. It got dark on the way home so I called for a D/F steer direct over the mountains to Aghione to save on fuel. They gave me a heading and sometime later we saw some lights below. We continued on assuming that we had seen lights on the West Coast of Corsica. What we didn't know was that the steer was to Bastia, in the north instead of what I'd requested. So, we were soon out over open water again and I had no idea where we were.

"About that time my wingman, Carrol Feather, checked in with, 'Red Leader I have about 5 minutes of fuel left.' I got on the radio again and asked for an emergency steer to the nearest base. They came right back with a steer but we couldn't see anything. I asked for a searchlight and, sure enough, they turned one on at the base at Borgo. We started to let down and I told Feather to leave formation and land and, 'make it on the first pass.' He went on down and we saw him on final approach, and then he picked up his gear and went around. I really started to sweat but he made it around and, as he was slowing down on his landing roll his right engine quit. Then as he tried to turn off the runway his left quit. I heaved a sigh of relief and the rest of us landed. As we went to Ops to report to base we were down safely I told Feather, 'Jeez, you really had me sweating there for a while', and he replied, 'Oh, I wasn't worried, I knew you'd take care of me.' So much for blind faith."

Escape photo, Warren Campbell, for Southern France invasion. (Campbell)

Flying from Corsica consisted of what by now were standard invasion duties, i.e., beachhead patrols and dive-bombing/strafing attacks inland as directed by the beachhead controller or as planned and flown directly from Corsica. The beachhead controller was code named "Baby", and patrol areas were designated as fruits,i.e., we were directed to patrol Grape sector at angels 10 or Apple sector at angels 5 meaning Grape at 10,000 feet altitude or Apple at 5,000. This would mean flying what amounted to a holding pattern covering perhaps 10 or 15 miles parallel with the coast at the specified altitude from there to be vectored to a specific target inland or to be on the lookout for anything untoward in the area. In flying these patterns one had to be careful not to overfly the U. S. Navy. They (the Navy) seemed to have a practice of firing at any plane that came into range of their ships. On August 14th, the day before the invasion, we dive-bombed a target identified as German Army headquarters seven miles north of Avignon. Results were very good with several hits on the main buildings. Southern France was defended by the German Nineteenth Army comprising seven infantry divisions and the 11th Panzer Division.

The Luftwaffe was noticeably absent during the invasion. On D-Day two groups of Me.109s were encountered. One flight of the 71st on its way to dive bomb, jettisoned their bombs and broke into a group of eleven. Robert Longworth shot one down, the rest fled. A 27th flight met a second group of six and Tom Maloney bagged two.

Major James Morehead, the leader of the second Cragmore flight, gives a good account of the engagement:

"Tom (Maloney) and I were on a dive-bombing mission over Southern France. Tom had a flight of four P-38s with 500 lb. bombs and I had a flight of four slung with eight more 500-pounders. Suddenly seven Me.109s were spotted diving on us. We turned into them meeting them head-on. I had been searching for our target below and hadn't seen

Chow time, Corsica. (Bertoglio)

them in time to toggle our bombs. Fortunately, no bombs were hit by German bullets and we dumped our ordnance and joined the fight. I got on the tail of a 109 and was gaining on him when he bailed out. I had not fired a shot but the reason for his bailout was, that he knew his 109 was no match for the American P-38 at low altitude both in speed and maneuverability and a few hundred feet off the ground where he was unable to dive away, he knew he was trapped. Germany accused its fighter pilots of bailing out of fights toward the end of the war as morale was low. Here was an example.

"I quickly turned to another target and followed him in a long dive that wound up flying down the gorge of the great Rhone river canyon - Europe's little Grand Canyon. Great towering walls rose up on both sides and I was perhaps 500 yards behind this 109 when suddenly from another arm of the canyon, out popped two more 109s directly in front of me in perfect range of perhaps 100 yards. I was centering my circle and dot sight when the old admonition of 'check your tail' came through, borne on the instinct of years of combat and previous bitter experience of wounded airplanes; the rattle of metal hail on aluminum; the sound of popcorn cooking off on your armor plate. A quick glance in my rear view mirror and damn! My finest opportunity shot. Three fine targets just off my nose, a full load of ammunition, plenty of gas and all I needed was a few seconds. But I didn't have them. The instant I jerked back on the stick and my faithful P-38 shot straight up alongside those cavernous walls of the canyon, I saw the wings and nose of the 109 light up with gunfire. He was close but he was late by 1/10 of a second.

"As I shot straight up in the sky out of the canyon my wingman, who had joined the rat-race from some distance, shot the 109 down. Fire burst from its wing roots and, as I continued around in my loop of the canyon floor (fortunately my loop alignment was true and I came out of it between the canyon walls) the German plane crashed against the walls of the great canyon in a fiery splash. Memories such as those of the smoking parts of the plane, engine, propeller blades, wheels, wingtips, perhaps arms and legs emerging from the fiery splash and bounding down the walls of the canyon will fade only when life fades."

Tom Maloney, who saw the action from too great a distance to be a part of it-although he tried, says Longworth caught the last Me.109 just as it started to follow Major Morehead up into the loop. He exploded him in a straight up position for what Tom describes as the prettiest victory he ever saw.

The invasion was costly for the 1st Fighter Group. Lieutenant Harold Kline, a 71st pilot, failed to return on 18 August and Lieutenant Lauren Erickson a Petdog pilot, went down the next day. Lieutenant Robert Taylor went down near St. Maurice and Lieutenant Walter St. John, both 94th, went down in flames on August 20th.

On the 19th eight P-38s from the 94th and 27th were led by Major Ed LaClare. They strafed a train west of Avignon. The train was loaded with ammunition and headed for the front. The strafing resulted in five large explosions. As a result either of the explosions or ground fire or both Dick Arrowsmith had to crash land on fire not far from the target area. Walter Gonring made it back to Corsica in a badly damaged P-38 with one engine out, no hydraulic system, one wheel locked down and no radio. He crash landed at Cape Calvi, Corsica, his ship a total loss.

CaptainTom Maloney, a 27th squadron pilot and one of the 1st Group's top scorers with 8 victories, had his plane badly damaged. Both engines lost oil pressure and he was forced to ditch in the ocean just off the French coast. Major Pope, the 27th flight leader, circled over Maloney to the limit of his fuel and then landed on a beachhead crash strip. Maloney remained afloat as it was too late in the day for a Dumbo rescue. The group sent out a search flight the next day, however it was unable to spot Captain Maloney who had ditched his aircraft five or six miles off a shore still under German control.

Captain Ed LaClare, 94th Squadron, attacks a locomotive (L. to R.) in the gun camera sequence above, then what appears to be trucks towing artillery below. The mission took place in August 1944 over Southern France. (Ed LaClare)

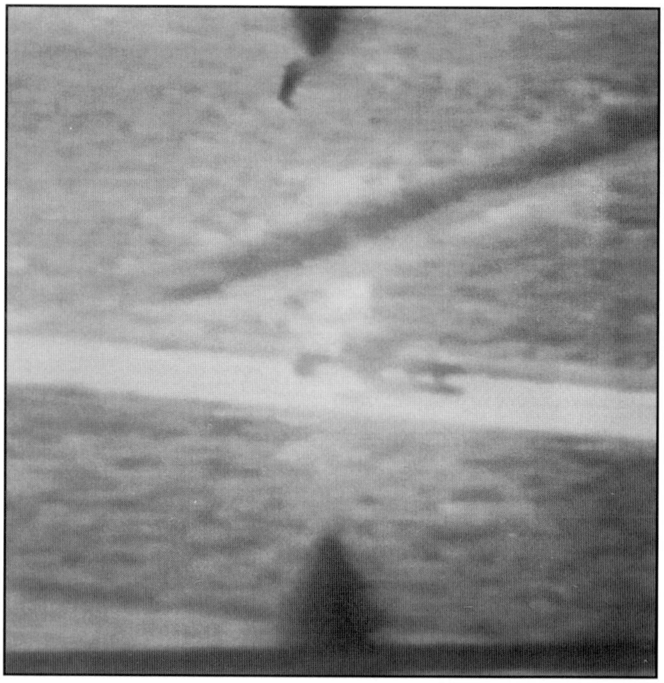

Captain Tom Maloney, 27th Sq. shown here at Foggia in May 1944, scored his seventh and eighth victories over Southern France in August, then was downed and seriously wounded. (Maloney)

After scrambling out of his now submersible P-38 he made it into a saucer sized dinghy and some hours later drifted to shore at a point near the Rhone river. Once on the beach, after dark, he made his way cautiously inland. After walking forty to fifty feet he heard a click followed by a powerful explosion under his feet. He had, of course, stepped on a land mine. The explosion drove steel shards through both feet and lacerated his legs with flying shrapnel. After lying semi-conscious for three days he began moving toward what he hoped was water, progressing by pulling himself backward in a sitting position. After about five days of this he reached a marshy area where he was able to construct a raft of sorts. Using this means of transportation he was able to reach a small French settlement. A group of men rigged a bench type car seat in the back of an old truck and started for a French hospital, however Tom's pain was so intense that he couldn't stand it. The men then carried him some three miles to the hospital. The hospital was poorly equipped and could do little for him, however, they did manage to contact the Americans about his condition. U.S. Army medics picked him up during his second night in the French hospital. After a stop at the 118th Station Hospital he was flown to an Army hospital in Naples where the first decision was to amputate both legs. Maloney implored 1st Group Commander, Colonel Richards, to intercede to try to save his legs. By whatever intercession, there was no amputation and Maloney was flown to the USA in November 1944 for further surgery. He remained bedfast until September 1945 at which time he was able to take a few steps with the aid of crutches.

Dick Arrowsmith was more fortunate. His ship caught fire, so he bellied in getting some burns in the process. After evading for three days he was picked up by a French farm hand. He lived with the French family for a week and picked up some interesting culinary information. Breakfast was black bread and ersatz coffee. The main meal, late afternoon, was usually vegetable stew except on lucky days. Dick and the man of the family went hunting every day; weapons were rocks and snares. They set the snares and beat the bushes back toward the set snares. Dick describes the feast:

> "A bird the size of a sparrow lit on a branch near us. Yves, my friend, threw a rock and (unbelievably) killed it - lucky day - meat in the stew pot. The sparrow went into the stew pot sans feathers only-feet, head, guts intact. At dinner that day the little bird was ceremoniously laid out on a small carving board. The head was removed and offered to me the honored guest. I declined having a little difficulty keeping breakfast down. Yves, the head of the household, took the little thing by the beak, popped it into his mouth and, with one sucking motion, cleaned that skull shiny. Wine was served with the evening meal."

After a week Dick was picked up by the French Resistance who took him through the lines thence back to Corsica where he rejoined the squadron. The August 19th mission was indeed costly as so many were but, as Ed LaClare put it, what better contribution to ending the war than by taking out a train loaded with ammunition and headed for the front.

Major Morehead, then serving as group Operations Officer made a similar observation:

> "As Operations Officer staging out of Corsica we received word of a number of German dive-bombers massed on an airfield near the front lines. We were directed to destroy them. Our group commander ordered the mission planned for strafing at tree top level. Now at tree top level the pilot flies as low as he can get, right down among the trees and between houses and barns. You pop up over the airfield and pick a target if you are lucky enough to be in line with it. You try for surprise by not coming in sight until you are there and you stay low among the trees after you pass the airfield so the gunners have nothing to shoot at.
>
> "Well those dive-bombers held the potential to decimate so many of our troops that I recommended that we come in at 5,000 feet so we could see and select our targets well ahead of arrival. Each pilot could pick a target, align his sights and have plenty of time to destroy one and maybe two or three. The C.O. said,'No we don't! I'm not going to get my boys killed going in at that altitude.' I said, 'Colonel Bob, we may lose a few of our men but those dive-bombers could take out a few hundred of our infantry troops and those boys

are just as dear to their mother's as we are.' I had backed away on other occasions, but I stayed and fought on this one and the C.O. finally gave in and let me plan the high approach, but he said he would hold me responsible. As it turned out the plan paid off but not as expected. As we approached the field at 5,000 feet we found that the Germans had evacuated their planes and pulled them back. If we had gone in at low level, the German gunners would have worked us over and we would have had nothing to shoot at."

Overall in the period August 10th to 18th the 1st and 14th together lost a total of twenty-five planes and a good many of their pilots. Nor were P-38s the only casualties in the air operation. One afternoon as we were taxiing out for takeoff the 14th Group B-17 supply plane was landing. This was a "retired" Fortress used to ferry supplies, mail and sometimes people. For some reason the pilot decided to go around. Something happened, perhaps the cargo shifted. At any rate the B-17 pulled up very steeply, stalled directly over the field and dived straight in. The plane just blew apart when it hit scattering pieces in all directions. Departing P-38s were literally trapped nose to tail on the taxiway. One B-17 wheel assembly careened across the area and wiped out the landing gear of a P-38 trapped in line. Fortunately nothing else caught fire and the mission proceeded more or less on schedule after debris was cleared allowing runway access.

The landings in Southern France took place on August 15th on the beaches between Nice and Toulon along with an airborne assault behind the beaches. Resistance was comparatively light and the American troops advanced rapidly. Almost too rapidly in some cases. We were at one point ordered by the beach controller to strafe tanks at a crossroads. We had just started our first strafing pass when someone shouted, "My God they're ours!" And they were indeed our tanks well ahead of their scheduled position. By the end of the month of August the ground forces had advanced nearly 180 miles and were approaching the city of Lyon.

In connection with ANVIL, Major Morehead recalls an incident in Corsica that gave a staff officer grey hairs:

"Sometimes the stress of planning and directing operations was greater than that of flying the toughest missions. While supporting the invasion of Southern France we would fly from dawn to dusk. We were operating out of the Island of Corsica on dirt strips bulldozed out of thick underbrush in a setting of low hills. The Germans were sneaking in on our troops with Dorniers and Junker dive bombers just at dusk. We were trying to catch them at a late hour which forced us into a night landing. One night we were returning to the darkened island over a dark sea with a faint moon over the horizon. I found the faint row of lights that looked more like a row of weak stars in a black sky than an airport runway. My flight and number two flight landed OK but the number nine man crashed on the runway. What came next was an Operations Officer's nightmare. Seven planes, low on fuel and no alternate fields to turn to. That left seven pilots to bail out or force land in hilly, brushy country at night. P-38s were going down in all directions and there was nothing we could do about it since we had no heavy equipment to drag the crashed plane off the strip. Five P-38s and two pilots were lost. Miraculously the others walked away. "

B-17 crash, Corsica. (Karpowicz)

Both P-38 groups departed Corsica for their Italian home bases on August 18th. During its eight days on the island flew a total of seventy-seven combat missions comprising 582 sorties. They carried a total of 191-1,000 pound bombs and 162-500 pound bombs. The group dive-bombed railroad bridges, highway tunnels, airdromes, and gun positions and strafed motor convoys, locomotives, rail cars and other military installations. Following is a list of known results, although there was often little opportunity to observe damage:

	Destroyed	Damaged
Motor Trucks	63	83
Locomotives	10	5
Freight Cars	36	25
Railroad & Highway Bridges	6	6
Power Stations	2	1

Once again statistics tell only a partial story. The group's P-38s set fire to many buildings and hangars, damaged tracks and road junctions, radar stations, oil tanks and, in short, did our damndest to hamper the ability of the Germans to resist the invasion, and we were successful. Even in leaving, however, the 27th suffered an unfortunate accident. Taking off to return to Foggia, through some miscalculation, Lieutenant Galen Rhoades and Henry Inge collided at low altitude, too low to recover or bail out and both were killed.

15

CLOSING THE RING

On the 31st of August 1944 Russian Armies overran the Ploesti oil complex and the city of Bucharest thus ending a campaign which started in April 1944. During the period April to August 1944 the 1st Fighter Group reported eighteen pilots missing or killed in action over that target area. The group flew a total of nineteen bomber escort missions, one mission escorting the 82nd in its dive-bombing raid and one strafing mission over the Ploesti area. These trips to the Ploesti refineries and the Bucharest area required an average long-range flight of some 1,200 miles round trip meaning a flying time of five and one half to seven hours, either at very high altitude or very low altitude. During this period the Fifteenth Air Force lost a total of 223 heavy bombers damaged in Ploesti operations. There is no figure readily available for total fighter losses.

An immediate benefit from the fall of Eastern Rumania was the mission flown on August 31st. We escorted B-17s to Popesti, just outside Bucharest, to begin the repatriation of nearly 1,000 Allied airmen held prisoner in the area. Six 1st Fighter pilots were among the returnees including Lieutenant Harry Noone, Edward Ulrich and Herbert Foor of the 27th; Lieutenants Ralph Hisey and Rudolph Jansi of the 71st and Raymond Geyman of the 94th.

As September began the 1st was lacking targets, and were also short on airplanes after the Corsican adventure. In a repeat of earlier travels there were two trips to England during the month to pick up more P-38s. For the most part these were new ships but the timing also coincided with the Eighth Air Force decision to convert most of their fighter groups to the P-51 aircraft.

On September 11th the following were dispatched to England for new aircraft: LaClare, Cake, Morris, Mullins, Gonring and Pregmon. The trip was one very cold, all night flight in the midsection of a B-17, destination Burtonwood just outside Liverpool. On arrival we were told the planes were not ready! Return in a week!!! Good news like that doesn't come very often, and within minutes we were on a train to London where we:

a. went sightseeing
b. nearly froze in our suntan uniforms and
c. got scolded by the MP's for being out of uniform.

Evidently no one in England ever wore suntans. We were told that summer was usually the first week in August (old joke), but probably not far from the truth for those accustomed to the heat of southern Italy. With the new P-38s we made the return trip non-stop carrying such exotic items as a gallon of coke syrup. Later in the month a second contingent made the trip to England. This group included Harris, Driggs, Clifford and Lindley. Once again transportation was via B-17 this time with a stop in Paris that Everett Lindley swears was an emergency.

The trips to England were made necessary by another shortage of aircraft and this is an appropriate time to mention the 94th Service Squadron. This squadron (the 94th designation is coincidental) was attached to the 1st Fighter Group in October 1943. Prior to that time the group was serviced by a variety of service squadrons some not as well equipped as the individual group squadrons. Beginning in October 1943 with the assignment of the 94th, 1st Fighter group maintenance record was exceptional. Between October 1943 and November 1944 the Service Squadron, performing major maintenance and overhaul, repaired 126 P-38s and one B-17. The sheet metal shop manufactured inboard camera mounts for all P-38s, manufactured and installed fuel tank guards to protect flaps when belly tanks were dropped and the litany goes on. Suffice to say that the squadron had much to do with the high efficiency record enjoyed by the group during the Corsica operation

Prior to the England trips Dave Raner, a 94th Fighter Squadron pilot, made a most unusual excursion, to Poltava, Russia to return a P-38, although this time

Col. Robert Richard, C.O. 1st Ftr. Grp. (Cake)

Lt. Dave Raner, who journeyed to Russia. Salsola, December 1945. (Raner)

not a new one. The shuttle missions which both the Eighth and Fifteenth Air Forces staged to Russia went on sporadically during the summer of 1944. Poltava was one of the Russian host airfields. The 1st Fighter Group was not involved in the shuttle missions. How Dave Raner got involved was a mystery to everyone including Dave.

He was one of six pilots sent to Poltava to return some P-38s which had been damaged and repaired during the shuttle missions. These were probably 14th or 82nd Group planes. As he describes the trip they flew from Foggia to Malta to Marble Arch to Lybia then to Cairo. Due to bad luck they missed a lot of connections and had to stay at the renowned Shepheard's Hotel in Cairo for a week. From Cairo they flew by C-46 to Teheran and via modified B-24 to Poltava. At Poltava the Russians demanded three to four days advance filing for a local area flight plan. The plane Dave flew back was repaired but barely; the ailerons stuck up above each wing like inverted flaps and he had to crank the gear up, but he was in no mood to wait for further repairs.

On one of the last shuttle missions the 20th Fighter Group of the Eighth Air Force flying P-51s landed at Salsola prior to returning to England. The leader of the 79th Squadron, 20th Group turned out to be Jack Ilfrey, ex 94th, who after a period of time in the USA, had been assigned to the 20th. Neither Raner nor the 20th pilots were overly pleased with the Russians. They described them as truly bastards.

September really represented a turning point in the war as it involved the 1st Fighter Group and indeed the entire Fifteenth Air Force. September became the first month when all three P-38 groups reported no enemy aircraft encounters - sightings yes - encounters no. The P-51 groups in all claimed only twelve enemy aircraft for the month.

With the two England trips the group spent a good part of the month equipping itself with new or reasonably new planes. Actually much of the activity during September was directed at the Balkans to harass and interfere with the Germans' retreat and escape routes. The heavies flew only twenty days and sixteen of those were directed against Balkan targets. Only three targets were in Germany and three more in Austria or Czechoslovakia.

A 1st Fighter Group pre-mission briefing. The target appears to be in Southern Germany, possibly Ragensburg. (Air Force)

Debriefing after mission, Salsola. (Bertoglio)

One organizational change, indirectly affecting the group, took place in September. The Fifteenth Fighter Command was created to include the existing 306th Fighter Wing, including only the P-51 groups and a new 305th Fighter Wing for the P-38 groups. September also saw the return to the 94th Squadron of Lieutenant Raymond Maxfield. He had been hit by flak and went down west of Bologna on June 5, 1944.

As October began, the weather took a decided turn for the worse. Too early for winter but the bombers operated only on sixteen days during the month and then, really in force, on fewer yet. Fairly warned by those who had experienced one Italian winter, preparations were started. The walkways among tents and buildings were formalized so to speak, by laying crushed stone obtained locally. We had had some experience within the group with tufa block construction and now building began in earnest. Tufa is a soft limestone that can be cut with a machete or even sawed. We positioned the location of the proposed huts or shacks then hired Italian laborers to do the construction. Wages were anything from cigarettes to clothes (socks, shirts, underwear, etc). Construction went quickly except for the roofing material.

L - R Pasteris, Furr and Clevenger, 27th Sq., working on new Allison engines, Salsola, 1944. (Bob Share)

Tufa blocks for winter shack construction, Italian laborers, Salsola, Fall 1944. (Karpowicz)

Normally that would have been belly tank crates, but a sharp eyed 94th pilot named Bill Dougherty located some corrugated metal sheets stacked (and guarded) in a storage area outside Foggia. An expedition went off one night in a weapons carrier and Dougherty talked his way past the guards.

He actually told them that a Colonel Athletesfoot wanted this stuff in Ancona by morning (the Germans still controlled Ancona!). The soon to be roofing material was trucked back to the squadron and hidden until;

a. weathering dulled its bright reflective surface and,

b. the furor over the loss subsided.

It did make excellent roofing. Floors were tile for the most part, again obtained locally via the barter system. Doors were constructed of wood, belly tank crates and windows were Plexiglas. Inside furnishings were handmade, wooden for the most part, some more elaborate than others depending on the carpentry skills of the inhabitants. Furnishings usually consisted of a desk, a bedframe to hold an air mattress and including lower shelves; closets and the by now standard one-half oil drum stove to burn 100 octane gasoline. The gas was contained in an elevated belly tank positioned outside the shack and piped in via copper tubing. All in all a very cozy arrangement.

Generally the shacks were four man, the same four originally in one of the tents. Dismantling the tents was an eye opening exercise. When we pulled up the PSP flooring we uncovered whole colonies of rat holes. There were a few attempts to shoot the rats using our .45 caliber pistols. This often after several medicinal

Tufa block shack interior, Dick Arrowsmith's quarters 94th squadron prepared for winter. (Air Force)

alcohol and grapefruit juice cocktails, which put the survival odds very much in favor of the rats.

The long-range escort missions continued through October when the weather permitted. On some occasions the weather in Italy permitted the whole mission to take off cnly to encounter an impenetrable wall of cloud in Northern Italy especially around the Alps. Then the whole train of bombers and fighters would have to turn around and fly back to their bases with nothing accomplished. However, here and there we did get through the cloud banks to attack Munich, Brux, Vienna, Brno in Czechoslovakia, Regensberg and similar targets. During October the group recorded only fourteen missions; eight were bomber escort, three strafing raids in Czechoslovakia, Hungary and northeast Italy, one photo escort and one C-47 escort to evacuate prisoners from Bucharest.

On October 14th following an escort mission to Bleckhammer, Germany the squadron strafed targets of opportunity on the return. Howard Dale was hit while strafing, caught fire, bailed out and was captured. On this same mission Bill Dougherty landed at Foggia Main after his hydraulic line was severed by anti-aircraft fire. And once again we had a training accident. Lieutenant Robert Jurgensen crashed and was killed on a local orientation flight. Almost invariably these local crashes involved new, inexperienced pilots. The odd thing is that not one of these fellows in this long string of local crashes bailed out, leaving the cause of the crash unknown or at least uncertain.

On October 20th the 71st Squadron had a very costly encounter with anti-aircraft fire without ever reaching the target area. The group assignment was to escort B-17s of the 5th Bomb Wing attacking oil targets in the vicinity of Ragensburg, Germany. The 1st Group, with Petdog leading, Springcap low left and Cragmore high right, encountered intense, accurate flak just south of Bologna, Italy. Cragmore Red, White and Green flights broke right and Blue fight left to avoid the flak. When Red flight was some miles northeast of the city Red #1 and #3 were hit. Red leader called Petdog leader that he had two on single-engine and was taking the squadron back to base. About this time Red #1 and White #4 collided in mid-air. The flying debris from this collision evidently hit Red #2 and in a few seconds he went into a spin and went in. There were no chutes. Blue leader began to escort Red #3, still on single engine, back to base. Green flight joined them. By the time the Bologna flak storm was behind them both Springcap and Petdog were out of sight and impossible to overtake. Lieutenants Olsen and Twedt collided; Lieutenant Wisner spun in and Captain Elliott crashlanded at Duress, Italy but was uninjured.[1]

The weather was somewhat better in November and flying activity picked up accordingly. There were a dozen bomber escort missions together with three strafing/dive-bombing missions in Yugoslavia and one escort/strafing mission to the Vienna area. Early in November near Munich two Me.163s were sighted but not encountered. In October, extending into November, Fifteenth Air Force gave more than passing attention to targets in Yugoslavia. Heavy bombers, dive-bombing and strafing attacks were intended to assist the Russian advances by interfering with the Wehrmacht as much as possible. Bridges, tunnels, convoys, rail systems, all were given our best attention.

On November 7th Major Bucky Harris led the group on a strafing raid attacking enemy convoys in northern Yugoslavia. This was followed by Captain Ed LaClare leading the 94th and 1st Group on a bomber escort mission to Vienna followed by strafing attacks on targets of opportunity south of Vienna, principally convoys and rail traffic. On November 16th Lieutenant Chester Heien led the 94th and group on a strafing attack against German truck convoys in the area southwest of and leading to Sarajevo, Yugoslavia. Dozens of trucks were either destroyed or damaged together with flak positions and troop casualties. The

This P-38J-15 was the mount of Herschell Baird 94th pilot, (in jeep). Names on the nose are BUCKY "N", SUZY B. The left engine cowling is labeled LINDA. (Baird)

Germans were prepared for this attack. Antiaircraft weapons were positioned along the highways and in armored carriers in the convoys.

Weather was increasingly becoming a factor as the month wore on. On this raid the weather was 9/10 overcast en route. A hole was found over the target so the P-38s got under the overcast for strafing, however, low lying clouds made it impossible to reach Sarajevo. We lost two on this raid, Evert Johnson, a Petdog pilot, was last seen in a spin with right wing and engine on fire. Edward Steffani, a 71st pilot, went down in the target area with a third pilot landing a totally disabled P-38 at Salsola.

On November 15th Colonel Arthur Agan assumed command of the 1st Fighter Group relieving Colonel Richard who had led the unit since September 1943.

The group's first verifiable encounter with a German jet fighter occurred on November 26, 1944. Lieutenant Royal Nyby was leading a flight of three Springcap P-38s; number four man had returned early. Everett Lindley was flying in the #3 position with Guy Thomas in the #2 spot. The flight was returning from a photo escort mission to Munich and encountered the Me. 262 near Innsbruck at an altitude of 26,000 feet. With Thomas lagging but attempting to get back into position, the 262 made a pass from the stern. Lindley called a break left. As the P-38s were about 90 degrees around, the 262 chandelled over them climbing into the haze. Thomas had disappeared. Neither Nyby nor Lindley was able to see what had happened to him.

Lt. Robert Bostwick, 94th tent area, Salsola, July 1944. (Bostwick)

In a post war exchange between Everett Lindley and the Me. 262 pilot, Rudolph Zinner, it was learned that Lieutenant Thomas was indeed shot down. He survived briefly in a German hospital but died of his wounds on December 4, 1944. He had flown only one or two missions.

The month of December was, in many respects, a continuation of November except that field conditions continued to deteriorate. Heavy rains made the camp areas, taxiways and even the landing strip very muddy. During the month nine heavy bomber escort missions and two strafing raids were flown.

The first on December 6th concentrated on communications targets in the Vienna area. The 27th destroyed one He. 111 on this mission. So many people were involved that an individual claimant was not determined.

On the last day of December Captain Bob Bostwick led the 94th and group in a fighter sweep in an area bounded by Munich, Regensburg and Passau, Germany. Destroyed were six enemy aircraft on the ground, two rail station houses, five locomotives, one power station, a radar station and numerous freight cars.

During December the accident/loss rate accelerated. Early in the month Lieutenant Louis Driggs and Richard Sykes both 94th pilots, had difficulties on a photo escort mission. Driggs, for unknown reasons, was injured in a bailout over friendly territory in Italy. Sykes ran short of oxygen, let down, ran into several enemy aircraft, was well shot up and had to bail out near Russian lines south of Budapest. Sykes, a tall, blond fellow, was nearly mistaken for a German. Fortunately, he was able to convince his captors otherwise. After a brief hospitalization Sykes returned to the squadron to find most of his cigarette hoard had disappeared. For that day and time Dick Sykes was an anomaly, a non-smoker who nonetheless hoarded his cigarettes for barter material.

Returning from rest camp at Capri Lieutenant Anthony Kainrad and Norberg Willig, the latter an armament officer, were killed in the crash of a B-24; the cause, weather. Lieutenant James Fairhurst was hospitalized with a badly injured leg the result of an accidental gun shot wound. The gun (a .45 caliber pistol) discharged as a hunting party was clambering into a jeep.

While Jim was hospitalized in Foggia and things were fresh in mind he jotted down air combat phrases that were then commonplace but quickly fade from memory. Here are a few:

- Bogies, 3 o'clock high!
- FLAK!
- Break Right!!
- Springcap prepare to drop bellytanks!
- Get! Him! Off! My! Ass!!!
- Spacebar from Springcap
- I'm hit! I'm leaving this crate!
- Bombers 12 o'clock low

A 94th Squadron P-38L-1 at Salsola, Italy in the winter of 1944-45. This was John Mullins' plane, DOTTY LOU, Serial No. 44-24746. (John Mullins via Karpowicz)

* They're '51's!
* 60 gallons in my left, 50 gallons in my right
* On one lung, going with the bombers!
* Lieutenant McEwen's call in the early dawn "Hubba, Hubba, gang don't forget your escape kits, you can't tell a 109 from a 51 without it."

In mid-December the field at Salsola began more and more to resemble a small lake. Taxiing on mud slick PSP became particularly exciting. Any miscalculation would slide a P-38 sideways into deep mud or forward where a spinning prop would slide inexorably toward the tail assembly of the P-38 in front. In one trial by mud a four-wheel-drive Jeep was mired to the floor boards between the 94th Squadron area and group headquarters. So on December 17th the group began operating from the airfield at Lucera, the home of the 301st Bomb Group, located just to the west. Back to the air echelon type operation once again.

The year 1944 ended in snowy, muddy Salsola on an upbeat note overall. The air war in particular was certainly going our way despite the problems with winter weather. There was even talk that some of the pilots would not complete their 50 missions in this theater but the entire group would be transferred to the Pacific in anticipation of the invasion of Japan. A disquieting thought but only one of the usual bag of rumors floating around. Countering this was a bangup Christmas party involving just about everyone and thoroughly enjoyed by all. For Christmas dinner we had turkey with all the trimmings, even ice cream for desert - manna from heaven. The new year arrived with a near blizzard bringing snow and a driving wind that discouraged much celebrating.

One of the occasional difficulties associated with the long-range, high altitude bomber escort missions, particularly after one extra mug of coffee was, "answering nature's call," to use the vernacular. When the need to urinate could not be put off the P-38 was equipped to handle the situation, or at least the cockpit design people thought it was. A hard rubber funnel was fitted under the seat with a hose leading down and out the bottom of the cockpit. One simply voided into the funnel and that took care of the problem. There were a number of things wrong with that scenario. First was clothing. The P-38 cockpit was cold. At 25,000 to 30,000 feet it was extremely cold. In defense we wore several layers of clothing. On top of the clothing we wore flight suits, fleece lined jackets and, in some cases, electric flying suit, plus a .45 caliber pistol and a large knife. Secondly, once installed in the cockpit we added a Mae West, a parachute and seat and safety harness belts and of course, helmet, oxygen mask and gloves. If one

Winter 1944-45, Salsola. (Karpowicz)

A 1st Group pilot in high altitude flight gear. (Eberhardt)

really, really had to go, the parachute straps had to be unfastened at a very minimum; various layers of clothing had to be unzipped and/or unbuttoned and, in accordance with the laws of nature, substances contract and shrink at low temperatures. The next maneuver is best described as akin to a worm being stretched out on a lawn following a rain. One is now, in theory, ready to use the relief tube. Since this operation was rarely carried out there is a very good chance that:

a. the tube will be kinked to stop flow
b. the tube will be cracked with obvious results, or
c. the outlet will be askew leading to a back air pressure which will indeed blow urine all over the cockpit where it will immediately freeze leading to instant near instrument conditions.

While all this interesting business is in progress one is also supposed to be flying formation, hoping that the anti-aircraft cannon crews are resting and praying also that enemy aircraft are busy elsewhere. The best solution - never drink much coffee before a mission.

1. This provides a good example of faulty memories. The author was on this mission flying as a Springcap flight leader. He remembers this incident as taking place on the return from Regensburg, with Cragmore Red leader and Red #2 being hit by flak high and 2 o'clock to his position. According to the records only the relative position turned out to be correct. So much for recollections.

Peel-up for landing, Salsola. (Bertoglio)

16

OF FURLOUGHS, VINCENZO AND YALTA

On January 8, 1945 we were bumped from the Lucera field. Another bomb group, also with weather problems, displaced us. At that point, just a year to the day since moving from Gioia del Colle to Salsola, the entire air echelon moved to the 82nd Fighter Group field at Vincenzo located southeast of Foggia. At Vincenzo the pilots moved in with the 82nd people; the ground crews set up their tents on the opposite side of the field. Things were crowded and cramped but the 82nd folks went out of their way to accommodate us.

January would hardly be termed a productive month. Only three bomber escort missions were carried out. As a group we flew 18 photo escort missions. On one of these, January 21st to Stuttgart, F/O Thomas Graffam, a 94th pilot, was shot down by anti-aircraft fire near Venice, Italy. Graffam was flying his first or second mission and was apparently having both oxygen and mechanical difficulties. He continued to fly straight and level, directly over the outskirts of Venice. German anti-aircraft fire tracked and hit him. His damaged plane gradually spiralled down and nearly pancaked into the Adriatic just offshore. He seemed at first to be swimming but apparently died in the water. In buzzing the scene one of the two remaining P-38s (John Mullins' mission # 53) had an engine shot out by anti-aircraft fire from the shoreline. On the same date James Hutchison was reported missing on a photo escort mission to Prague. He was forced to go on single engine then bailed out in heavy overcast.

A handful of furloughs began to be offered to long serving non-coms. Margaret Anderson, whose husband Armand was a 71st mechanic, relates the tale of his good luck aided by his squadronmates:

> "Armand had a son nearly two years of age whom he had never seen, so the question of furlough was of major importance. One evening 'Andy' was off to visit a friend of his in a neighboring group; the friend was headed for the States. As Andy was leaving his tent he told his buddies that he expected to have his name drawn in the lottery for furloughs and would they please see that his bag was packed. On his return from the farewell bash Andy found everyone in his tent snoring. When he got to his corner of the tent, after groping in the dark to find his cot, he stumbled over something. He lit a match and saw his barracks bag all packed and tied up tight. 'Those jokers,' he thought, 'they think they'll get me all excited. I'd never be that lucky.' So, smiling to himself he went to sleep and dreamed of home.
>
> "Andy woke to his buddies standing around shaking him and saying, 'Hey Andy, better get

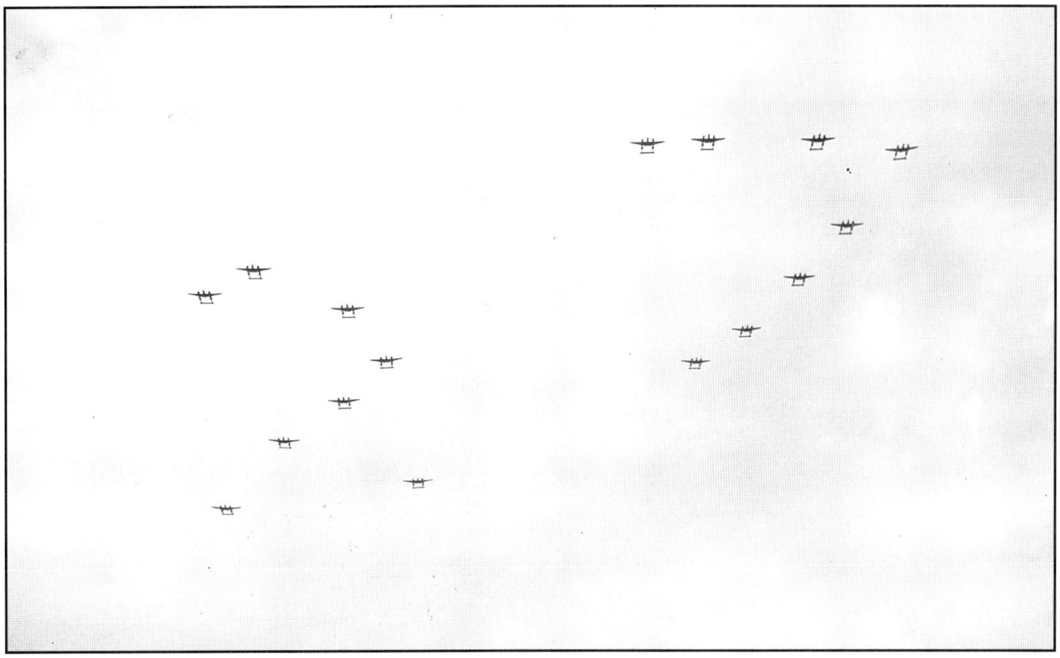

27th Squadron - who else? (Bob Share)

27th Squadron NCO club, Salsola, 1944. L - R unknown, Finkelstein, Meacham, Christianson. (Bob Share)

going or you'll miss your flight.' Still thinking it a joke he turned over and put up a hand to signal them to go away. After rolling him out they convinced him to go look at the company bulletin board. There it was in black and white: Anderson, Armand T. He had made it, home on furlough which lasted until April. Years later at a 1st Fighter Group reunion Andy was reminded of that lucky day when he had gotten to go home. Only then did he learn that the name drawn that evening had been that of his buddy Ted Sheldon, but Ted had asked to let Andy take his place. 'I'm not married and he's got a baby he's never seen,' he said. Andy made it home just a few days after our son's second birthday"

February was another hiatus month, at least it seemed so in the beginning. However, the 1st had been selected to provide escort for the principals attending the Yalta Conference. Termed Operation ARGONAUT, this presented the group with an interesting and different challenge. Planning actually began with the assignment on January 16, 1945 to provide continuous fighter cover for both sea and air transport of American and British officials en route to and on return from a conference at Yalta in the Russian Crimea. Coverage would commence at Gibraltar, extend all the way across the Mediterranean to Yalta and return.

Fighter cover had to be available at the proper times and places throughout the Mediterranean area to assure continuous daylight escort of participants. Specific information on both naval and air transport was difficult to obtain because of the need for secrecy and the impracticability of long-range scheduling. Winter weather in the Mediterranean area was an additional variable.

To assure continuous escort with backups for weather and other variables, units of the 1st staged at various locations and leapfrogged across the Mediterranean. Staging points included Gibraltar, Oran, Algiers, Tunis, Malta, Athens, Alexandria, Cairo and the Crimea. Russ Michels and about a dozen others went to Gibraltar to service the P-38s at that end of the Mediterranean. The morning of February 11th in Athens typified the weather problems. Severe frost and ground fog had coated the P-38s with a thick layer of clear ice which built up even more heavily on the props and wing edges during preflighting. The ice was chipped off by all hands using hammers, wrenches and other similar tools. Despite this exercise and the ground fog, the P-38s took off on schedule. Among those escorted that morning were the U.S. Chief of Naval Operations and the U.S. Army Chief of Staff flying in separate C-54s.

The Yalta operation covered 16,000 miles and included three continents. For a while no one knew quite who or what was involved until one day the following was heard in the clear over the radio, " Mr. Eden wishes to thank the LIghtning pilots for their splendid escort." At that time Anthony Eden was the British Foreign Secretary. Colonel Agan was thanked by both President Roosevelt and Prime Minister Churchill.

The Yalta adventure was completed on February 19th and on the 21st the group returned to a dried out Salsola base complete with a new theater made out of tufa block. The new theater was said to be the largest structure of its kind in the Fifteenth Air Force and should be a magnet for USO and stage shows. While the Yalta operation was taking place the engineers, for all practical purposes, rebuilt the runway at Salsola. The weather cooperated in this, but fate intervened in other ways. Early in February the 27th Squadron officers club caught fire when gasoline from one of the stoves overran. The roof went as did much of the furniture, however, rebuilding began almost immediately.

The stage was now set for the final two months of the war. Once again, as it did at Mateur, the group would move out just after a major rebuilding effort. Whoever followed us into Salsola would get a well constructed runway, some comfortable shacks, nice clubs and the best theater in the Fifteenth Air Force.

17

FINAL HORROR AND ACCOUNTING

The squadrons had barely settled into their old stamping grounds at Salsola when the final move of the war began; this time to the shores of Lake Lesina near the Adriatic, northeast of San Severo. A move of only a few miles. The relocation took place between March 8 and 16, 1945. The Lake Lesina location had a better runway and generally better ground conditions. Despite the move, this was a busy flying period for all three squadrons; one of the busiest in unit history. Missions were flown without a break from March 8th through the 26th. Sixteen heavy bomber escorts were flown plus seven dive-bomb/strafing missions to Austria, Germany and Yugoslavia.

On March 9th on a dive-bombing mission to the marshalling yards at Knittelfeld, Germany Lieutenant John Felsinger, a 94th pilot, was killed in action. Aircraft dove from about 6,500 feet through a hole in 9/10 cloud cover pulling out from 1,000 to 2,500 feet. Lieutenant Felsinger dove at a bridge over the Mur River, hit it with his bomb and pulled out low. He was hit immediately by anti-aircraft, fire rolled twice and crashed in the target area.

By this time all three squadrons were equipped with the final combat modification of the Lightning, the P-38L. One innovation the L included came in very handy on a photo escort mission to the Regensburg area on March 14th. An Me. 262 made a pass at the escort, shots were fired but no damage was sustained or inflicted. Tail-end charlie on this mission had good reason to be pleased with the rear warning radar installation. He did not see the 262 as it closed on his tail but evaded successfully upon signal of the radar warning light.

The group got a bit of public relations in mid month. Several writers visited the 1st Fighter Group in order to get background information for magazine articles and books. Several of the visitors were given piggyback rides in P-38s, visited with pilots and attended an interrogation of pilots returning from a mission. Colonel Agan gave them a briefing describing the planning and execution of missions. The 1st was recommended to the visitors by the public relations office at MAAF as being the most interesting group (naturally).

Formation view of number 65, a 94th ship. It was a P-38L-1. (Bertoglio)

Another interesting visitor during March was Mr. Harold Tittmann, assistant to Myron Taylor (President Roosevelt's representative to the Vatican). Tittmann's visit was particularly appropriate inasmuch as he had been a pilot in the 94th Squadron during WW I, participating in the battles of Toul and Chateau Thierry. Lieutenant Colonel Bucky Harris brought Tittmann to the group from Rome in a piggyback P-38, a ride which he thoroughly enjoyed. Major Ed LaClare took him back to Rome in the wing B-25.

On the 25th of March the group strafed the area between Regensberg and Nuremberg. Principal targets were railroad stock including a train protected by a pair of flak cars that mounted two so-called "bandstands," 20mm or slightly heavier caliber braces of four guns each; a total of eight per car. Each stand of guns was surrounded by a circular shield of armor plate, the whole resembling a cheese box or bandstand.

On March 31st the mission was a fighter sweep in the Vienna area. Colonel Arthur Agan, group C.O., was flying as Springcap Red Leader although the 71st Squadron was leading the group. Beyond the Alps the squadrons became separated partly because of weather. The 94th and 71st were flying between cloud layers at about 13,000 feet in the vicinity of Lake Balaton. Both squadrons dove through a hole in the cloud cover. Flying at an altitude of 2,500 to 3,000 feet the P-38s ran into a hail of anti-aircraft fire. Flak came up from scores of gun positions on the hillsides and from motor transport. This position was very close to the Russian frontlines. Colonel Agan and Lieutenant R.G. Gillen were both hit. Colonel Agan, with his plane on fire, headed east toward the Russian lines, but the fire intensified and he was forced to bail out literally over the German front line trenches and was taken prisoner. Gillen was missing. The 71st also ran into intense flak and small arms fire and two P-38s were hit; one managed to return to base, the other, Lieutenant Crawford was last seen on single engine headed for the Russian lines. The 27th shot down a FW 190 and an He.111 and destroyed a large amount of railroad stock as did the 71st, however, the 27th lost two pilots, Lieutenants Hendricks and Coury, the latter from anti-aircraft fire and the former during the Heinkel chase. Overall the group lost five that day. The 94th, after getting out of flak alley, spotted a cluster of unidentified aircraft, dropped belly tanks and gave chase. They turned out to be Russian Yaks and A-20's. At that point the fuel supply mandated a course for home base.

Col. Arthur Agan C.O. 1st Ftr. Grp. (Cake)

The group lost ten pilots during the month including Arthur Agan the group Commander. Additionally we lost Lieutenants Edgar Coury, Kary Hendricks, Norman Crawford, R.G. Gillen, J.H. Tappan, Steven Dostal, John Felsinger, Major Paul Ash and Lieutenant Colonel Charles Thaxton, recently posted as Deputy Group C.O. Who survived was questionable at the time, although Thaxton was seen to have parachuted and reached the ground safely.

While several of the group pilots were taken prisoner during the war, we do not have much information about their experiences in durance vile. One exception is Cy Nolen, a Springcap pilot shot down in Northern Italy in 1944. Cy was imprisoned in, he believes, Stalag Luft #3 located east of Berlin. He describes what he terms Super Bowl 1944 but first they had to prepare the field beginning with removal of tree stumps without benefit of tools such as picks and shovels.

With the field prepared a volley ball league was started with each prison barracks fielding a team. That led to a football league, touch only since many of the men had only partial clothing let alone equipment. Nolen's barracks had the benefit of a former professional baseball pitcher who could really throw a football. Since touch football is a passing game their barracks won but was immediately challenged to a playoff by "all stars" made up of all the other barracks. Cy Nolen's barracks, led by their star passer "Bullet" Bruce Kilmer did their darndest but wound up on the short end of a 3 to 0 score. Even the Germans were impressed although they had little idea of the game.

Shortly after this, with the Russian artillery lighting up the sky to the east Nolen and many of his fellow inmates were transferred to a prison in the vicinity of Nurenburg. This involved a forced march from a Saturday morning to the following Monday afternoon. By then, late winter or early spring 1945, Germany was breaking apart and Nolen was able to escape with a friend and make his way to the advancing American Army.

The Nurenburg prison was close to a railroad yard which was bombed by the British and the Americans with some regularity. Shown an escape route by two fellow prisoners Nolen and his buddy Dick Mart decided to escape (their second attempt) using a British (night) air raid as cover. The German guards and guard dogs were in bomb shelters, the POW's in fox holes near their barracks. Using a loose board from the latrine Nolen and Mart ran for the fence. They slithered under the barbed wire, using the board to bridge the concertina wire strung between fences. With the help of German flares they evaded the grenade trip wires and made it out of the prison. This accomplished they ran from a barking guard dog and promptly fell into a large pit. Exiting the pit, their next obstacle was a wide, cold river with a bridge and, of course, a guard on the bridge. With patience they waited for the guard to move on and then walked across the bridge and kept on going, heading west, moving at night, sleeping during the day. Food, or lack of it was a problem until they discovered a field of planted potatoes.

Finally they knew they were close to the front lines by the number of sentries they encountered or avoided encountering. One needed a certain "sixth sense" to feel the presence of a sentry ahead somewhere in the dark. At dawn one day they saw an American tank coming along the road. A German 88 began firing from just above them and they were soon in the middle of a brief fire fight. This over, they found themselves behind American lines but had some difficulty flagging down a ride. Who was going to pick up two ragged looking individuals walking down the road in old French Army overcoats. After a typical American hail to a passing Jeep, namely, "Hey, you sons o'bitches!", the Jeep stopped, they piled in and were taken first to a field kitchen where they had a meal of eggs and bread and promptly fell asleep, finally safe and secure.

April 1945 arrived as the last full month of the war in Europe, and it was also the busiest month in the history of the group. Missions flown reflect the changing nature of the war with fourteen dive-bombing missions, twelve bomber escorts, (some of these latter two concurrent), two fighter sweep/strafing attacks and twenty-nine photo escorts. Armed reconnaissance missions including both dive-bombing and strafing were flown on three days totalling ten missions. That contrasts with April 1944 when eighteen bomber escort missions were flown together with only one strafing mission. In some respects the group was flying ground support missions (albeit longer range) however, this time with planes and pilots to accomplish the task.

Inclement weather, particularly around the mountains, continued to be a problem. On an April 6th dive-bombing mission to southern Germany the group got into very severe cloud conditions in Northern Italy. Flights became badly split in an effort to get through heavy snow and black clouds, all this with mountains below. Finally everyone turned around, jettisoned bombs over the Adriatic, and went home. The mission was repeated the next day. The P-38s penetrated into Germany only to find solid cloud cover north of the Alps. After letting down to 2,000 feet and finding no bottom, the mission repeated the jettisoning and return to base. Finally on the third try, April 8th, the weather cleared enough for the plan to proceed although there was some disagreement on which bridge was the target and how successful the bombing. The Seefeld bridge strike was repeated on the 9th. The bridge, with a tunnel to the southeast, was difficult to line up for a bomb run due to mountainous terrain. An Me. 262 was encountered on this mission without consequence to either side.

After an escort mission on the 14th the group began dividing its efforts once again. On the 15th one force of thirty-two P-38s escorted a formation of B-17s attacking German troop concentrations around Bologna, Italy in cooperation with the U.S. Fifth Army offensive in Northern Italy. The trip, led by Captain Walter Gonring, was a milk run in terms of opposition for the fighters, however, the effects of the heavy bombers on the German troop concentrations was catastrophic and did much to aid the Allied armies advance in Northern Italy.

The second force of P-38s strafed rail lines in Southern Germany, specifically in the area of Regensburg-Salzburg-Munich. On these fighter sweep missions the group split up into squadrons to cover separate areas. The 71st departed Lake Chiem and strafed the area northwest toward Munich; the 94th targeted the area from Regensburg south toward Landshut, and the 27th strafed from Regensburg southeast parallel with the Danube toward Passau. The 27th and 71st stationed one flight, at least, as top cover. The 94th declined to use top cover because of low clouds. Targets were plentiful as indicated by the following claims: eleven locomotives destroyed, ten damaged, nineteen freight cars damaged, four tank cars, two troop cars, one flak car, three single engine aircraft on flat cars, three river barges, six passenger (RR) cars all damaged and one FW 190 downed. However, the 1st Fighter Group lost five pilots on this mission including all three squadron leaders.

At this stage of the war the German armies were being compressed. The Russians were advancing from the east, the Americans and British from the south and west. The Germans, their firepower for the most part intact, were being squeezed and attacking this mass, bristling with guns, was like poking a stick at a hornet's nest. Anti-aircraft fire came from every roadside and hillside. The sides of freight cars fell away to disclose heavy machine guns and anti-aircraft cannon manned for the most part by very skilled gunners. Lord knows they had had enough experience.

Flying into the haze and smoke beneath a 1,000 foot overcast the strafing P-38s ran into a hail of machine gun and small cannon fire on April 15, 1945. Of the five pilots shot down that day only one survived; Captain Everett Lindley, the 94th Squadron leader and mission leader. Captain Chester Trout the 27th squadron leader, was hit by flak just west of Passau and crashed in flames. Major Joseph Elliot, the 71st leader, also went down the victim of anti-aircraft fire. Lieutenant Edward Brickley was hit and bailed out, Lieutenant George McGrew was hit by flak and his P-38 crashed and exploded. Everett Lindley's story describes the chain of events leading to a bail out and the chaos existing in Germany as the end of the war drew near. It is also one of the rare detailed reports of a downed pilot's experiences:

"I was assigned to lead the 94th and the 1st Fighter group on a low level strike from Munich to Regensberg to Linz. All hell broke loose right in the middle of my strafing run at the Landshut marshalling yard. I was firing at a long line of freight cars, in a 45 degree diving approach when, out of nowhere, at about 700 feet, came a concentrated volley of twenty to thirty orange balls from 40mm flak guns. Too late to turn and fire into the nest of flak guns, I was forced to pull out. However, before I could completely extricate myself from this mess, I was struck by at least six of those shells. The first one ricocheted off the two inch glass windscreen, yet it was powerful enough, even with this glancing blow, to pulverize that glass. It broke loose on the inside and that exploding glass hit me in the left shoulder. I could still move my arm but did not want to look at my shoulder. As I pulled out I could hear those orange balls striking the aircraft. Two struck the right engine, one the wing and two penetrated beneath the gondola making contact with the armor plated bucket seat. I could both hear and feel those malicious last two. I knew that these 'armour piercing incendiary' shells could cause troublesome fire eventually.

"As I climbed to about 1,500 feet flames started to appear around the right engine. I tried diving to extinguish the fire but this seemed to increase the burning. I then started the procedure to eliminate the number two engine. I feathered the prop and cut off all supply to that engine, i.e., gasoline and electrical power. This was to no avail and the fire worsened. About this time heavy smoke permeated with flame came from the area of my feet and into the cockpit. Because my goggles tended to restrict my peripheral vision I did not wear them while strafing. As a consequence the cockpit flames started to burn my face and eyes between the oxygen mask and my helmet.

"When the cockpit filled with dense smoke, and I could no longer see the instruments, I was forced to release the emergency panel on top. This action cleared the cockpit of smoke momentarily and I could see again, but it also sucked the flames from the burning shells in the gondola. Meanwhile, as I glanced at the right engine, not only was it on fire but most of the right wing was engulfed in flames.

"I had completed my single-engine procedure except for trimming. I was holding hard left rudder when the cockpit became a mass of fire! Because of the flames now emanating from the right side and the cockpit and realizing the possibility of the aircraft exploding, I rolled down the window and evacuated the cockpit on the left side. I stepped out into the slipstream of the number one engine which was at near 95% power. This movement pinned me back against the rear cockpit edge. The cushion effect resulting from wearing our parachute on our back, thus sitting on our dinghies, caused a jamming condition from which I could not break loose. Not having had the time to trim the aircraft for single engine operation, the P-38 was now in a right spin. As I fought to break clear the long headset cord was still connected and I could hear someone yelling, 'Red leader, bail out! Bail out Red Leader!' While I could not move, I did realize the ground was close.

"The urgent call to bail out came once again and I made one last effort to break clear of the aircraft. Why this remarkable P-38 did not explode with all the fire and fuel aboard I shall never know. In any case, with my right hand on the ripcord, I made, with all my might, one last desperate effort to break loose and succeeded. From a hot noisy environment I found myself in a delightfully cool and quiet situation. I looked up at the chute and I could see houses, trees and farmland and, as I looked down at my feet I saw only sky. Obviously I was upside down. I remembered in training they told us when you are oscillating in a parachute jump, manipulate the shrouds to stabilize your descent. I got my hands only halfway up to the shrouds when I slammed into the ground snapping in two the fibula bone in my right leg. If I had landed on my spine it would have been the end. Moreover, as I hit, my head impacted the surface and I was knocked momentarily unconscious and the chute was all over me. I had landed in the village of Stenging just north of Muhldorf, Germany.

"When I came to, my squadron mates were strafing around me to provide an opportunity for escape. As I freed myself from the chute I watched two enemy riflemen approaching. Noting that my squadron had, by this time departed, I scrambled for the woods dragging a broken leg behind me. The forest wasn't much of a forest! It was simply a narrow band of trees planted between two levels of farmland to prevent soil erosion. I soon, therefore, ran out of woods and into another open area. In this action I had aroused a spring lair of deer to include a buck, doe and fawn, that ran into an open area. What a telltale indication the deer provided for anyone wanting to know my whereabouts.

"I then retreated back into the limited woodland and tried to swallow a wadded up "flimsy sheet" (a sheet with mission details). I could not do this because my mouth had no saliva. I quickly buried the classified document. A German civilian, who I supposed was part of a posse, noticed the frightened deer and continued walking until he walked right by me. He said something like, 'Oh,Oh,' but did not seem belligerent. It was obvious to me that he would inform others of my exact location.

"This is apparently what happened because soon afterwards I was completely surrounded by hostile forces. I drew and cocked my .45 automatic and pointed it at the area where many seemed to have congregated. I remembered thinking that I had worn this weapon on 43 missions and it had its purpose. It was to afford some protection in one's effort to avoid capture. Here now was the reason to implement its use.

"Someone yelled, 'Hands up!', and I could hear weapons being put in their ready to fire state. Then others approached from my rear. I immediately turned their way with my .45 pointed at them. God, what was I doing with this macho behavior against some 30 enemy? Another again yelled, 'Hands up!' and I stood there, confronting them in what seemed an eternity. I think I was ready to die knowing I would get a few of them before they got me. It was an electrifying situation and I know in retrospect that if I had allowed another second of time to pass they would have opened fire. I guess it was 'discretion being the better part of valor', belatedly, that took over and I decided to toss my .45 to the ground. Immediately I was surrounded and, with my hands up ordered to follow them out of the woods. As I proceeded to the lower level of ground I could see and hear women fighting over my silk parachute. I was put in a vehicle and, under guard, driven from Stengling to the town of Muhldorf.

"Unknown to me the dramatic confrontation with my captors was only beginning. I remembered what the intelligence officers told us to do if we were shot down. To escape, contact the Partisans in Yugoslavia, the Russians on the eastern front, the Catholics in Germany, they said. And, if captured, give the enemy only your name, rank and serial number. Ultimately you would, they said, be transferred to a P.O.W. camp. Being shot down then never seemed to me to be such a big deal. But if you knew what could really happen, despite what they told you, some airmen quite possibly would have refused to fly combat! (Catch 22)."

Lt. Everett Lindley (right) shown here with 94th C.O., Bucky Harris, was downed on April 15, 1945, within days of war's end. (E. Lindley)

Taken to a public building in Muhldorf by armed civilians, Lindley was alternately taunted and beaten by his guards and some local citizens. No effort was made to tend his wounds.

"A vehicle finally arrived with two German Storm Troopers. They were Obersturmbahnfuhrer Colonel Mathias Zeirhut and his aide Sergeant Yohann Gilch. They took command of the situation and ordered me into their Volkswagen. I was directed to the front seat with the Sergeant as driver and the Colonel took the back seat. We drove swiftly out of Muhldorf and into the countryside. I assumed that these people were salvaging me from the wrath of the German civilians and were transporting me to a P.O.W. camp. Soon the 'bug' came to a stop in the boondocks and an irate Colonel tapped the back of my head with his pistol and told me, 'Grouse mitt you!' I obediently got out of the car with my hands up and faced him. I noticed when the vehicle stopped the driver got out and proceeded to the area of the left front headlight. With his back to me he seemed to be fiddling around his fly and my first thought was that he needed to answer a call of nature. In actuality he was loading his pistol. Meanwhile when Colonel Zeirhut departed the back seat I got a good look at his eyes. They were glossy and he appeared to have a nervous demeanor. He walked around me with pistol in hand and, when he arrived at the front of the vehicle, he raised his right arm and fired! I had sensed his intention when I noticed his eyes and I winced when the shot was fired. Somehow it missed me and Sergeant Gilch pointed his weapon at me. Zeirhut then walked completely around the car and came up from the rear to within six feet of me and fired again. This time, in spite of my ducking in anticipation, the round struck me in the left shoulder. It felt like a straight arm one receives in football because it pierced the bone area and spun me around and into the ground. I knew then 'the jig was up.' Rather than be shot to death there on the side of the road I elected to run for it. I ran across a pasture like a broken field runner, hobbling on a broken leg. Both Zeirhut and Gilch were firing at me and I could hear bullets whizzing by my head and ricocheting off the ground. There was pain from my right leg but it was insignificant. It's true, in desperate circumstances like this your life memories such as Christmases, food and women you have known, do pass through your mind's eye. And I remember as I ran looking at the grass and singling out a living blade and envying it. I can still, to this day, see blood pumping out of my shoulder through my flight suit at every accelerated heartbeat.. Oh, God, how do I live through this?

"I also felt a great thirst when I noticed a small river ahead of me. My mouth was foaming white, dry saliva if you can believe this. The only moisture I could get was from my still bleeding nose. I was again ready to die if they would allow me first a drink. But as I ran zigzagging and favoring my right leg, my good left leg stiffened up and I finally tripped over myself. I fell to the ground totally out of sync, when the approaching Zeirhut kicked me in the head and yelled once again, 'Grouse mitt you!' For the first time I was really angered and now ready to kill! I learned after WW II from the War Crimes trial records that a young German lad who had witnessed the action described above, ran into the house of Joseph Kettner, a Wehrmacht Sergeant home on leave and shouted 'They are killing the American!' The whole family, apparently having lunch after attending church on this Sunday in Catholic Bavaria, departed the house to observe. They congregated behind a substantial hedge, therefore I did not see them.

"It was at this time that I felt a profound desire to live. Realizing that I had survived to this point, I must have felt a tremendous positive condition overtaking me and wanted only to be spared. As I say, Zierhut kicked me in the head as I lay on the ground. I was for the first time in my life really angry and realized that only drastic behavior was in order if I was to live. When I stood up he pushed me toward the river, banked by trees, and again I had this terrific desire for drinking water. If they would only allow me a drink of water, I was, I think, prepared to die!

"However, as he pushed me I knew that they were not interested in my thirst. I then decided to make my move. I was truly on the horns of a dilemma. It was either die or take on these two armed and murderous Germans. An awkward situation with unpleasant alternatives, but now the choice was clear. As he shoved me again with his left arm, holding the pistol in his right hand I reached over with my right hand and grabbed his right wrist. With my left hand I jammed the muzzle of the pistol. I did not concern myself with the possibility of another gunshot wound. I don't know where the strength came from but I picked up Zierhut with this grip and swung him around knocking Gilch to the ground. As I fought for his pistol he finally let go and I pointed it at them as they ran away. I was going to shoot them when I noticed a group of people, including children, at whom they were running, standing behind a dense hedge. Apparently Zeirhut had seen them as he shot at me and thought they would provide some protection. I don't know where one gets the presence of mind in this kind of crisis to restrain one's self, but I elected not to fire for fear of hitting someone in the group.

"I saw a small bridge that spanned the river and, with haste, headed for it. There was only open area around me. With the pistol I would get some of them, I felt, so I backed away toward the bridge. When I got to the high point of the bridge, with no place to hide and feeling totally naked, I stopped to gather my wits. Directly an elderly German farmer approached me beckoning me to go with him, because,as he said, I was 'kronk' (sick), meaning I was bleeding. I raised the pistol at him and he backed off. Although I felt he was genuine in his appeal, he knew I was an injured animal and therefore dangerous. He tried another time, begging me to 'komen sie her' and failed. On the third effort he was accompanied by his son, a Wehrmacht Sergeant, Joseph Kettner. I finally acquiesced when I could see from their eyes that they were really sympathetic and now joined them. As we walked off the bridge, I noticed a person behind some bushes getting off a bicycle removing a telescopic equipped rifle from his back and I called out his position to the Kettners. The father screamed some words at this person and he quickly replaced his rifle on his back and cycled away. This further confirmed my confidence in them.

"You must realize that I felt completely lost and desperately seeking any indication of salvation. This still was a critically dangerous situation. When we arrived in their yard by the house, all the people were there. They consisted of old folks, children and some in my age group. As I stared into their eyes I could tell they were opposed to what Zeirhut and Gilch had done. I can't recall what the villains were doing during this time frame. They seemed to be in the background and hiding. I think they hoped the Kettners would disarm me and turn me over to them. My immediate attention was drawn to the Kettners. While they exhibited great courage to attempt to assist me as I leveled a pistol at them, they were violating a major international war principle; i.e., aiding an armed enemy!

"Suddenly young Kettner yelled something in German and everyone except his father ran into the house. I assumed that the Kettners would now attempt to kill me and that they did not want the others to see. I rapidly raised my pistol at these two remaining Germans. They backed off and begged me not to shoot. Again, from their eyes, I could perceive that they were genuinely expressing their true feelings. Directly all those who had disappeared came running out of the house with hot water, bandages and medicine. One lovely woman, who was perhaps sixteen years of age, came over and looked to me and looked compassionately into my eyes. Gosh, I melted and, if this was a trick, I would have been willing again to die! However, they surrounded me with help to stop the bleeding. Sergeant Kettner rolled back my flying suit and with hot compacts, swathed my shoulder and sprinkled sulphur powder on my wound. Then he applied a tourniquet and bandages. Meanwhile, his father, who sat next to me, urged me to give up the gun. 'Give me sie gun, comrade,' he said repeatedly. Although I considered that this might be another trick and realized the worst, I finally gave Herr Kettner the pistol in view of the aforementioned humanitarianism. Nonetheless, I was still cynical and immediately expected, once more, that he would kill me despite my belief in human nature. One has to remember that these people were from a different culture unlike any I had known. How was I to know to what extent they would respond? After I gave him the weapon I looked at him expecting the worst and Oh God, I was again spared!! He put the gun under his shirt and said 'Danka!' Also, I noticed that Zeirhut and Gilch were now being harassed by their fellow Germans. These good people were now shaking their fists in the faces of the villains. Again, after the war, I learned that during this melee, Kettner had dispatched the same young lad who had earlier alerted them, to the local Luftwaffe station for assistance.

"A Luftwaffe Hauptman (captain) finally arrived to mediate the argument between Sergeant Kettner and Colonel Zeirhut. Kettner stood at attention, red faced, while Zeirhut walked back and forth making his point to regain control of me. With Kettner were his family and neighbors, all standing in a line, who had witnessed this event. While the twenty minute debate for my fate was taking place, two of the largest police dogs I have ever seen approached me as directed, and sniffed at me. One order from their master and they would have torn me apart. True, in my unsettled state, even an unfriendly Chihuahua would have looked enormous!"

Rescued by the Luftwaffe, Lindley was taken to an air field where he saw FW 190s tucked away under trees in revetments and the runway was an autobahn. A Luftwaffe doctor gave him some modest medical attention, and in excellent English, pronounced Lindley, "a good sport."

Driven to a POW camp at Freising Lindley witnessed P-51s strafing en route. The camp doctor examined his wounds and applied some anti-tetanus serum which caused a reaction.

"I was placed in solitary confinement where for six days I was comatose and nearly died. I am sure that when the prison officials first saw me on arrival, they realized I was very ill and, therefore, placed me in solitary confinement, hopefully, to recuperate.

"The previous events that took place, any one of which could have been fatal, had a telling impact of my psyche. As for example, the near explosion of my P-38, the near failure of my bailout, confronting a large group of hostile Germans with a .45 pistol, beaten up by an enraged civilian, driven to the countryside by two StormTroopers and shot at, waiting for an open court to decide my fate—thus escaping Zierhut's demand to recapture me, and being allergic to anti-tetanus serum. These unnerving experiences had cost a lot mentally and physically and, with the shock of losing much blood, had weakened me to the point of not being able to resist the old-fashioned and possibly contaminated anti-tetanus serum.

"I did wake on the third or fourth day a bit hungry but went right back into unconsciousness. On the following Sunday I was removed from this lone cell and placed with two wonderful human beings. They were Peter Cook and Warren Williams. Peter was an RAF night bomber pilot and former medical student and would daily apply moist compresses to my leg and also massage it. He was a prince of a fellow. Warren was an American B-17 navigator who came from Buffalo, New York.

"I have spent some time, since WW II with the dentist Joseph L. Kettner in Munich. He told me that Zeirhut returned to their home during the evening of April 15, 1945 with a vengeance and a band of armed soldiers. They surrounded the house and opened fire hoping the Kettners would surrender. Since the Kettners expected retaliation they had evacuated the area. The parents of Joseph were located in the end and arrested. His father was subjected to severe electric shock treatment in Munich by the Gestapo and finally imprisoned at Dachau. His mother was released in due course, but Joseph had to desert the German army until the war ended. All this because Zeirhut was certain that the Kettners were Allied sympathizers. Sergeant Kettner had fought on both the Italian and Russian fronts.

Everett Lindley freed from POW camp enjoying a plate of fried eggs near Freising, Germany, 1945. (Everett Lindley)"

"As a result of my hair-raising adventure on April 15th I have come to love the Germans! I am sure the Kettners represented the majority, in peace and even in war, who were fair and humane. For them there was a point beyond which they could not go! I had a chance after the war to forgive, though certainly never to forget, the diabolical behavior of Zeirhut when I received a letter from his lawyer. He begged me to consider the plight of his client, who was wasting away in prison, and urged me to communicate with the War Department to seek a pardon for him. What he stated in his letter was very interesting. He said, 'You are home safe and sound and enjoying life.' And this did give me cause to realize how extremely lucky I was, and therefore, I wanted to ponder this opportunity to be merciful. When I discussed the matter with my father, a very biased man, I cannot possibly record here the language that he used! I also confided with some of my peers about my desire. However, they were unsympathetic. One colleague, for whom I had the greatest respect, informed me that people must pay for their wrongdoing if civilization is to advance and so he felt that the 'punishment fit the crime.' With that I waited for more correspondence but it never came."

Thus ends the saga of Captain Everett Lindley. He was liberated by Captain Fry of Virginia, C.O. of the 94th Reconnaissance Squadron, Patton's Third U.S. Army. His experience does raise a question about the fate of at least two of the other pilots lost on that April 15th raid. The two other squadron leaders, Captain Trout and Major Elliot, were seen to go straight in after being hit by anti-aircraft fire. However, their two wingmen, Lieutenants Ed Brickley and George McGrew were thought to have bailed out after being hit. Both were killed, circumstances unknown. What happened to Everett Lindley became a matter of concern to the post war War Crimes Division. A War Crimes Investigating Team pursued the case into 1947. At that time Everett was notified that Zierhut was sentenced to life inprisonment and Gilch to a term of ten years. The war crimes investigations verified and/or identified individuals who aided or did not aid Everett Lindley during his incarceration.

In the meantime, back at Lake Lesina, the war was rapidly drawing to a close. The missions continued, both of the dive-bombing/strafing variety and bomber escort, however, it was becoming more and more obvious that the Germans could not last much longer. The balance of the month was made up of missions to the Innsbruck area on April 20th, the Munich area on the 21st, the Po Valley on the 22nd and the Padova area on April 23rd where, once again, two out of the three squadron leaders and one flight leader were downed by flak while strafing.

1st Ftr. Grp. Base at Lake Lesina, spring 1945. (Karpowicz)

The mission to dive-bomb and strafe in the Padova/Verona area was intended to interfere with the retreating German ground armies. Captain Larry Reynolds, the 27th Squadron leader, and Captain John Hurst a Petdog flight leader were hit by flak and both bailed out, apparently successfully. Captain Clarence Knapp, the Cragmore leader was hit by intense anti-aircraft fire while flying at an altitude of only fifty feet. The aircraft did a half roll, struck the ground and exploded. In an interesting footnote to the 71st Squadron report, "The Group Air Inspector who participated on this mission, believes that the damage inflicted against the losses sustained does not make this type of mission practical in Northern Italy." A factor influencing this comment might have been the difficulty of locating targets combined with the intensity of anti-aircraft fire.

An unusual higher command situation arose in April as a result of Colonel Agan's loss. When Agan went down Lieutenant Colonel Milton Ashkins[1] assumed command. A few days later Lieutenant Colonel Charles Thaxton, who had been missing in action since early March, returned to the group and assumed command on April 11th. Thaxton had been sheltered by Yugoslav Partisans since parachuting from his plane. Later Thaxton had to be hospitalized at Bari and Ashkins once again took command of the 1st Fighter Group.

One more unusual note; the 1st Fighter Group became the tacit owners of the first two operational jet fighters in the European Theater. Operational is probably too strong a word because the two short range P-80s, dubbed the 33rd Air Force, couldn't possibly accomplish anything of an operational nature from that far south in Italy, assuming the powers that be would risk having such an aircraft shot at in the first place. Harry McConnell, referring to a story appearing in the July 1946 issue of AAF Review, notes that the P-80s (called Shooting Stars) were brought over early in April by a Wright Field contingent, "for testing under combat conditions in a remote location". The remote location was correct but combat conditions? Surely not. It seems pretty certain that few if any of our people flew these aircraft although Lieutenant James Franey recalls flying formation with one or both, which seems reasonable. They usually flew formation and carried out simulated combat maneuvers with either the P-38s or the P-51's. Major Ed LaClare, by then assigned to wing, recorded two flights in the P-80. There were the usual rumors linking the P-80s the 1st Fighter Group and the Pacific Theater of Operations. More to the point, the new jet fighters were obviously the key to the future and there was probably a long lineup of those anxious to fly one, though very few did. There are no records, at least readily available, to tell us what was to be accomplished by locating the two P-80s at Lesina. To be sure, they were in an operational area but aside from occasional flights in the local area not a great deal was accomplished. Nor are there records telling us what became of the aircraft at war's end, although we can be sure they were returned to the States. Surely these aircraft were not abandoned in the wilds of Italy as was so much of our equipment.

Late in April Norman Crawford, a Cragmore pilot, returned to the group. Crawford had been a flight leader on the March 31st mission strafing targets in the Vienna area, on the same mission where Agan was downed. Crawford's ship was hit by flak knocking out the left engine. He reached Pecs, Hungary where he crash landed, wheels up. He was picked up by a Russian Lieutenant and a Bulgarian soldier and taken to town where he was joined eventually by three B-24 crews.

After several days of poor food and accommodations, bad train service and Russian bureaucracy they were airlifted back to Italy on a C-47. Crawford returned to the 1st Group nearly a month after his crash landing.

P-80 at Lake Lesina. (Bertoglio)

Early May saw an end to hostilities in Europe. On May 6, 1945 five P-38s of the 27th squadron escorted two RAF Halifax aircraft to Yugoslavia to drop supplies in the mountains west of Celje. This was the last combat mission flown in WW II by the 1st Fighter Group. The first had been a scramble of two 94th Squadron P-38s from Ibsley, England on August 31, 1942 for bogeys over the North Sea.

VE-Day was in some respects anticlimactic for it had been apparent for weeks that Germany was on the verge of collapse. The Wehrmacht had been cut to pieces and isolated into pockets. The Luftwaffe had some planes but few pilots and almost no fuel. The major question in the group and indeed in the minds of most of the American personnel was "What next?" Some hints became apparent when the training program was started on May 7th. It consisted of dive-bombing and gunnery flight training together with ground school classes in Japanese aircraft recognition, lectures on Far East service and training films. Skeet shooting, physical training and link training were instituted for all concerned in anticipation of transfer to the Far East. During May most of the group personnel were started on their return to the United States ostensibly for transhipment to the Pacific. The atomic bomb changed all those plans and brought about the Japanese surrender. What would have happened to the men of the the 1st Fighter Group had an invasion of the Japanese home islands come about is pure conjecture and an unpleasant one at that. Fortunately for everyone it never came to pass.

And so it finally ended. From the defense of the West Coast in early 1942, to a remote outpost halfway up the boot of Italy in 1945. Three and one half years of emotions and experiences ranging from utter boredom, i.e., the Air Corps infamous hurry up and wait!, through routine duties, the quiet enjoyment of mail call, the satisfaction of a job well done, good friends, occasional sheer fright and, for some, the exhilaration of flying the P-38 Lightning. All of the latter counterpointed by good friends lost. We also became connoisseurs of English chow, North African sand and dust, (Tafaraoui..where the mud is thick and gooey) Italian mud, C and K rations, al fresco dining and tent living. We were introduced to Labrador mosquitoes, English marmalade, extended shipboard travel, palm trees in a desert oasis, the strangeness of Africa, and the desperate poverty of bombed out, fought over villages in Italy-in brief and to paraphrase:

> We're glad that we came but damned anxious to go,
> Give it back to the natives, we're ready to blow!

1. In a previous tour of combat in Alaska with the 54th Fighter Squadron Ashkins had downed a Zeke near Kiska.

P-80's of 1st Fighter Group in formation near Vesuvius, spring 1945. USAF Photographic Collection National Air & Space Museum, Smithsonian Institution.

EPILOGUE

For some, perhaps most, this was the end of a unique, never to be forgotten experience; for others the foundation of a successful career in the soon to be United States Air Force. For all of us our time in the 1st Fighter Group and its squadrons, the 27th, 71st and 94th will stand forever as an example of collective skills and courage when our country was badly in need of both. We came from all walks of life, mostly untrained and inexperienced, and, together, made up a group capable of and more than willing to tackle anything the enemy fielded be it bombs from a Ju. 88, shells from anti-aircraft cannon or machine gun bullets from a Messerschmitt or Focke Wulf. We gave much better than we got.

The summer of 1945 was occupied with the usual military routines. A pleasant surprise in mid-May was a visit by Colonel Arthur Agan. Shot down over Austria March 31, 1945, he parachuted from his burning aircraft and landed in German trenches only a short distance from the Russian lines. As a prisoner of senior rank he had acted for several weeks as commander of Allied airmen incarcerated at Moosburg. Although he had lost weight Colonel Agan appeared none the worse for his experience.

Rumors intensified during May. However, during the month there was a substantial reduction in personnel. All three squadron commanders departed as plans were discussed to celebrate the group's third overseas anniversary.

By pure chance, Captain Joe Szaflarski (center) meets two 27th Squadron pilots in Paris after VE-Day, Henry Smith (left) and Burt Weil. Joe had completed his 54 mission tour with the 27th and was flying with the Air Transport Command. Smith and Weil had just been released after nearly one and one-half years in POW Stalags. (Joe Szaflarski)

During June, activities focused on the training program for the Far East theater of war. It was assumed that most if not all would see service in the that theater before Japan was finally conquered. A number of education courses were made available and, as a part of recreation facilities, the Fifteenth Air Force instituted opportunities to overfly parts of Germany to see first hand the results of the strategic bombing campaign. Harvey McDaniel, one of the pilots remaining, recalls a swap of two each P-38s for two each P-51Ds; this trade with the 332nd Fighter Group, our geographic neighbors in the then segregated Air Corps.

Venice was added to the list of rest camps and many took advantage of the opportunity to visit the historic city. Cannes was also a popular spot with the bright sun and sand of the Cote D'Azur being particularly inviting. During June each squadron maintained its compliment of 27 pilots and overall numbers stabilized allowing some compression of facilities. The 27th and 71st moved into the Headquarters and 94th areas. By the end of June the group had a total of 29 P-38s for all three squadrons. There were three piggy-back's and one B-25; the latter used mostly to transport visitors and rest camp quotas. New P-38s, those with less than 100 hours, were flown to Capodichino for shipment to the Far East Theater. Those remaining were used for training. group personnel settled in and made themselves comfortable. There was no official word on status other than the group was in 'category 2' meaning it would be redeployed. When and to where remained unknown.

And so the summer progressed. The training program was maintained with emphasis on Japanese aircraft and Far East tactics and strategy. A topic of considerable interest was that dealing with jet propulsion and in particular the German jet and rocket propelled aircraft which were discovered during the survey of the Reich following the surrender of the Wehrmacht. Flying training totalled 20 hours per pilot per month including some night flights, the first time in at least a year for flying at night. All was accomplished without mishap with a single exception. 1st Lieutenant George McGinnis, a 27th pilot, was killed on July 26th when his P-38 went into a dive from 1500 feet and crashed near the Po River during a navigational training flight. Cause of the accident was not determined.

Life at the Lesina base continued as a routine of recreation and training stressing Japanese air tactics until mid-August and the Japanese surrender. At that point the group began to divide at least in its thinking, into those interested in the Air Corps as a career and those preferring to return to the United States. For those desiring the Air Corps as a career, duties to broaden their experience were assigned accordingly. Although the group continued to turn in its worn P-38s to the Air Depot a fairly well rounded program of flying was maintained. Morale was generally good although a number of the men were restless by this time and were looking forward to returning to the United States, and gradually they did.

Harvey McDaniel, a 94th pilot who was there to the end, tells us how the summer progressed:

> "In August we flew all remaining P-38s to Bari for salvage. This was heartbreaking to leave almost new P-38Ls where instruments, radios etc. were removed then the remains were crushed and sold. During August we acquired several war weary B-25s which we checked out on our own. These aircraft were then used to transport high ranking Army personnel from remote locations in Northern Italy to Naples for deployment to the USA. We made some hair raising flights into small Army dirt strips to pick up these Generals. I recall several times I had to buzz the strip to chase off sheep before I could land. Most of these flights were flown with no co-pilot as we had only three or four pilots remaining.
>
> "During August all buildings except for group headquarters, the Officers Club and our tent residences were sold and turned over to Italian business men for salvage. The Italians who purchased the buildings posted guards but, as I recall, people came from all directions some as far away as Northern Italy. They descended on the base, ran off the guards and proceeded to tear down and remove everything. We had to resort to pistols to keep them away from buildings and equipment that was not sold. Cy Williams, myself and the few other officers remaining packaged the squadron records and shipped them to the U.S. Archives in the States and deactivated the 94th squadron on September 18, 1945."

In mid-September the group, or what remained of it, relocated to Caserta, near Naples. On October 16, 1945 the 1st Fighter Group was deactivated, ending 36 years of continuous service to the United States.

The 1st Fighter Group was reactivated by decree in 1946. In July of that year the 412th Fighter Group at March Field was inactivated (on July 3rd). Its personnel and equipment were assigned to the 1st Fighter Group newly activated at March on the same date. In effect exit 412th, enter 1st. The three squadrons, the 27th, 71st and 94th were reactivated along with the group, all flying P-80 Shooting Stars. Ralph Schell recalls that when the 412th was redesignated the 1st he wound up in the 71st squadron. He says the early P-80s were the ultimate in simplicity but with a few problems such as no canopy ejection, no seat ejection and real problems in fuel control. The engines had to be pulled every five hours and sent to San Bernardino for tear down and inspection. All was not in vain however; the group made the cover of Life Magazine in 1946 as the only operational jet fighter outfit in the (then) Army Air Corps.

The United States Air Force was established in September 1947 and the 1st became the 1st Fighter Wing in August 1947. Over the next several years the 1st went through a variety of administrative and organizational changes as the Air Force coped with growth as well as the Korean War. The squadrons were dispersed separately and the Wing was actually inactivated in 1952. Reactivated again in 1955, it was once again inactivated in 1961.

The group formed its own aerial demonstration team in 1950. Dubbed the Sabre Dancers and composed of 27th Squadron pilots the unit reached its high point in 1950 when it performed before President Harry Truman and most of his Cabinet. During the next several years the 1st would be a part of three different air forces, three regional air defense commands, an air division and four major commands. Through the 1960's the squadrons were assigned separately and became scattered throughout the country. By 1969 the 27th was at Bangor, Maine assigned to the 1st Air Force; the 71st was a part of the Tenth Air Force stationed at Malmstrom AFB, Montana; the 94th was briefly at Osan, Korea then assigned to Wurtsmith, Michigan, formerly the 1st Pursuit training camp at Oscoda. The 1st Fighter Wing, with no units under its control, was transferred to Hamilton AFB, California. In effect the 1st Fighter Wing barely existed and that in name only.

The 1st Fighter Wing was reincarnated, literally, in 1970 due to the historical awareness of General William Momyer commander of the Tactical Air Command. General Momyer directed his staff to redesignate units in his command to retain designations of units which had a combat record dating from either WW I I or Korea. This policy led to the return of the 1st Fighter Wing to TAC in October 1970. The means were somewhat unusual and took place in May 1971. The commanders of the 45th, 46th and 47th Tactical Fighter Squadrons participated in a "shootout" at the Avon Park Gunnery Range on May 14, 1971. The command had decided that these squadrons would become the 27th, 71st and 94th Squadrons assigned to the 1st Wing. Lieutenant Colonel Martin of the 47th was high scorer, had first choice, and chose the 94th designation. Lieutenant Colonel Walsh was second and chose the 27th. This left the 45th squadron to become the 71st. In this way were the 1st Fighter Wing and the three original squadrons, the 27th, 71st and 94th reactivated thanks to General Momyer. The Wing was assigned to the Ninth Air Force that summer.

In 1975, after at least a year of deliberation, the Air Force decided that its most historic unit would become the first operational F-15 wing stationed at Langley. The Wing built during the following year to reach its authorized strength of seventy two aircraft in three twenty four plane squadrons. This was accomplished in 1976 for the 27th and 71st. The 94th reached the same status by the end of 1978 having been hindered somewhat by the Air Force need to develop a second F-15 wing in Europe-requiring experienced pilots from the 1st.

Despite occasional difficulties, i.e., a spare parts problem that made headlines in 1980 the 1st Fighter Wing made steady progress during the 1980's and in 1985 won acknowledgement as the premier F-15 Fighter Wing in the Tactical Air Command-a position it holds at this writing.

In a somewhat parallel development, late in the 1970's a few of the members of the 1st Fighter Group during WW II began to think about days of old and whatever happened to so and so. After some effort, phone calls and letters to old addresses the first reunion of the 1st Fighter Group former members took place in Torrey Pines, California in 1976. Forty six people attended and the Association was founded with Irvine Styer as the first president. The group has met on a biennial basis ever since and continues to do so. A newsletter edited by Irvine Styer, begun in the spring of 1983 has done much to hold the group together. In 1986 the Association dedicated a memorial at the Air Force Museum in Dayton, Ohio. In much the same fashion members of individual squadrons began to locate one another hold reunions to reminisce and enjoy each other's company. All came together to support the memorial construction and dedication over a several year period culminating in the 1986 dedication. This effort was headed by Charlie McCann of the 71st squadron. The memorial was designed by Warren Perryman of the 27th squadron. Warren's design won in a blind competition.

The present day 27th Squadron has its own unique memorial to its predecessor squadron in WW II. The squadron has, and will continue to have an aircraft named MALONEY'S PONY - at this writing, an F-15. The original MALONEY'S PONY was a P-38 assigned to Tom Maloney, 27th and 1st Fighter Group leading ace in WW II.

The airfield at Salsola, Italy visited in 1993, is a large plowed field in an area totally given over to agriculture. The 27th squadron mascot, Antonio Castriota also visited in 1993, is a successful merchant in Foggia. Foggia itself is a pleasant, prosperous community showing no vestige of its wartime devastation.

Fifty years after the fact the events of World War II stand out in the memories of those of us who experienced them. The Group memorial at the Air Force Museum, a plaque dedicated at the Air Force Academy and this history are tangible reminders of those events. Each of us has memories and memorabilia, less tangible but none the less real, of the events of that war. It may be trite to repeat that history forgotten is history destined

Tom Maloney in 1993 F-15 version of Maloney's Pony. (Maloney)

Monument to the 1st Fighter Group erected at the Air Force Museum, Wright-Patterson AFB, Dayton, Ohio. (Mullins)

to be repeated. The events of WW II will not be forgotten and pray God they will never be repeated.

It seems appropriate to paraphrase (by changing tenses) Sir Winston Churchill's words spoken at Harvard University in September 1943.

" How proud we ought to be, young and old, to have lived in this tremendous, thrilling, formative epoch in the human story, and how fortunate it was for the world that when these great trials came upon it, there was a generation that terror could not conquer and brutal violence could not enslave.

"Let all of us remember that we were on the stage of history and that whatever part we played, great or small, our conduct was scrutinized not only by history but by our own decendants. We did indeed rise to the full level of our duty and of our opportunity and we thank God for the spiritual rewards He has granted us for all forms of valiant and faithful service."

A P-38 and pilot come home to Salsola and a relieved crew chief. (Bertoglio)

APPENDIX A

1st FIGHTER GROUP LOCATIONS WWII

San Diego Naval Air Station
Dec. 9 1941
Los Angeles Regional Center (HQ Sqd)
Jan 1 '42 -May 20 '42
Mines Field Los Angeles (27th Sqd)
Dec. 30 '42-May 20 '42
Glendale Municipal Airport (71st Sqd)
Feb 12 '42 - May 20 '42
Long Beach Airport (94th Sqd)
Feb. 6 '42-May 20,'42
Kirton in Lindsey, England (94th)
June 10 '42
Goxhill (HQ, 27th ground & 71st)
June 10,
Ibsley , England, (71st & 94th)
Aug. 24, 1942
Wellington, England, (27th)
Aug. 26
Colerne (Bath) (27th)
Sept. 15,
Chivenor, Lands End, all squadrons
Nov. 6, 42
Tafaraoui, Algeria
Nov. 15 '42
Nouvion, Algeria
Nov. 21 '42
Youks les Bains, Algeria (94th)
Nov. 28 '42
(detachments from Maison Blanche)
Dec. 6-14, '42
Biskra, Tunisia
Dec. 14 '42
Ain m'Lilla, Tunisia (71st)
Jan 24-Feb. 15, '43
Chateau dun du Rhumel, Tunisia
Feb. 14 '43
Mateur, Tunisia
June 20 '43
detachments from Dittaino, Sicily
Sept. 6-18
Gambut, Libya
Oct. 4-12
Djedeida, Tunisia
Nov. 1 '43
Monserrato, Sardinia
Nov. 29 '43
Gioia del Colle, Italy
Dec. 10 '43
Salsola, Italy
Jan. 8, '44
detachments from Aghione, Corsica
Aug 10-18 '44
Vincenzo, Italy
Jan. 9-Feb. 18, '45
Salsola, Italy
Feb. 19 '45
Lesina, Italy
Mar. 16,1945
(Omits staging at Dow,Goose Bay, Greenland and Iceland)

From: **Combat Squadrons of the Air Force WWII**

APPENDIX B

ACES OF THE 1ST FIGHTER GROUP

Owens, Joel	5-1-1	27th Sqd.
Welch, Darrell	5-0-3	27th Sqd.
Wolford, John	5-0-2	27th Sqd.
Kennedy, Daniel	5-0-2	27th Sqd.
Mackay, John	6-1-0	27th Sqd.
Maloney, Tom	8-1-3	27th Sqd.
Miller, Armour	6-0-2	27th Sqd.
Tovrea, Philip	8-1-3	27th Sqd.
Sears, Meldrum	7-0-0	71st Sqd.
Wiseman, Lee	5-0-0	71st Sqd.
Fisher,Rodney	5-0-1	71st Sqd.
Morehead, James B	8 (7 in Pacific)	71st Sqd.
Hatch, Herbert	5-1.5-3	71st Sqd.
Ilfrey, Jack	6 (plus 2 with 8th AF)	94th Sqd.
Roberts, Newell	5-1-1	94th Sqd.
Donald Kienholz	6-0-1	94th Sqd.
Lathrope, Franklin	5-0-1	94th Sqd.
Lee, Richard	5-1-0	94th Sqd.
Miller, Everett	5-0-0	94th Sqd.

* Taken from USAF Historical Study # 85, Albert F. Simpson Historical Research Center, now the USAF Historical Research Center, and USAAF (Mediterranean Theater) Credits For The Destruction of Enemy Aircraft In Air-To-Air Combat In World War 2 compiled by Frank Olynyk. Numbers above indicate, from left, aircraft destroyed, probably destroyed and damaged. In addition to the pilots listed above there were a number pilots with one or more confirmed victories, together with probables and damaged, as well as enemy planes destroyed on the ground. These last are almost impossible to accumulate accurately at this writing due to incomplete records however, the totals are probably considerable. Tom Maloney of the 27th squadron, in addition to the above, destroyed 8 enemy aircraft on the ground and damaged 4 more.

In its 32 months of combat the 1st Fighter Group destroyed 440 enemy aircraft, probably destroyed 103.5 others and damaged an additional 245. This at a cost of 204 pilots killed or missing in action with an additional 28 killed in accidents. This does not include those pilots taken prisoner or those shot down who evaded.

Totals of enemy aircraft destroyed, probably destroyed or damaged, by squadron, is as follows:

RECORD OF AERIAL VICTORIES

	Destroyed	Probable	Damaged
Group Hq	1.33	1	9
27th Fighter Sq.	187	45	97
71st Fighter Sq.	115	25	58
94th Fighter Sq.	136.66	32.5	81
Totals	440	103.5	245

As noted above these totals do not include enemy aircraft destroyed or damaged on the ground. There is simply no way to arrive at accurate numbers for such categories but it seems safe to say that the above totals would be increased by at least 50 per cent. The US Air Force Historical Study 85, published in 1978, lists only enemy aircraft destroyed in the air. More recently, in 1987, Frank Olynyk compiled what he terms Victory Lists which includes enemy aircraft destroyed, probably destoyed and damaged. The information is compiled by theater. The Mediterranean Theater is List #6 entitled USAAF (Mediterranean Theater) Credits For The Destruction Of Enemy Aircraft In Air-To-Air Combat World War 2. The Olynyk listings are comprehensive and are given in several book sections. The first lists victories in chronological order, the second section lists claims by unit, the third lists claims by names; the last three sections list claim totals by date, unit and name. Neither Study 85 nor the Olynyk books include information on ground claims

APPENDIX C
1st FIGHTER GROUP INDIVIDUAL VICTORY CREDITS

Pilot victory credits are given alphabetically by squadron. Listing is the usual confirmed-probable-damaged tabulated from left to right. Data are from the Air Force Historical Study 85 and from USAAF Credits For The Destruction Of Enemy Aircraft In Air-To-Air Combat In World War II(Mediterranean Theater) by Frank J Olynyk.

27th Fighter Squadron

Capt. James Alford	2-0-1
Lt. Robert B. Anderson	0-0-1
Lt. Robert H. Anderson	0-0-2 (1st Ftr. Grp)
Lt William Arnold	1-0-1
Lt Robert Austin	0-0-2
Lt George Bancroft	2-1-0
Capt. Richard Barr	1-0-0
Lt Stuart Bennett	1-1-2
Lt Robert Britz	1-0-0
Lt Charles Brown	4-0-0
Lt Merle Brown	1-0-1
Lt Robert Burgoyne	0-1-0
Lt Gilbert Butler	2-1-0
Lt William Caughlin	3-1-4
Lt David Conn	2-0-0
Lt Richard Cooley	0-0-1
Lt Edwin Crane	2-1-3
Lt Warren Danielson	1-0-0
F.O. Donald Deisenroth	1-0-0
Lt Ronald Delaney	1-1-0
Lt Harry Dowd	2-0-0
Capt. Warren Eickmann	1-0-0
Lt George Eldred, Jr	2-1-1
Lt Duane Enyeart	2-1-2
Capt. Blair Erickson	0.166-0-0
Lt Joe Ewers	1-0-0
Lt Arden Ferren	3-0-0
Capt. Walter Flynn	0-0-1
Lt Alden Freng	0-0-1
Col Ralph Garman	0.33-0-1(1st Ftr.Grp.)
Lt Frank Gerry	3-0-2
Capt. Joseph Glenn	1-0-0
Lt Elmer Hartman	1-0-0
Lt Kary Hendrix	0.166-0-0
Lt Donald Hilgert	2-0-1
Lt Robert Hill	0-0-1
Lt Otto Hloucal	2-0-0
Lt Warren Holden	2-0-1
Capt. Richard Huber	4-2-1
Lt John Hurst	3-1-1
Lt John Husby	1-1-0
Lt Paul Janson	2-0-0
Lt William Johnson	2-0-1
Lt James Joye	1-0-0
Lt Daniel Kennedy	5-0-2
Lt. Othel Kilpatrick	1-0-0
Lt Jim Kuykendall	1-0-0
Lt Lester Laird	1-0-0
Lt Harold Lange, Jr	0-1-1
Lt Stanley Lau	4-0-2
Capt. Francis Lawson	2-0-1
Lt Harold Lienau	2-2-2
Lt James Lily	1-1-1
Lt Marcus Linn	1-1-4 (27 &37 Sqds.)
Lt Eldred Loder	3-0-0
Lt. Roderick Love	0-1.66-0
Lt. Samuel Lyons	1-0-0
Lt John MacKay	6-1-0
Capt. Thomas Maloney	8-1-3
Lt Francis Maron	1-0-1
Lt. William Martyn	1-0-0
Lt Frank McIntosh	3-0-2
Lt Robert McIntosh	1-0-1
Lt Herbert McQown	3-0.5-2
Lt. James Meikle	0-0-1
Capt. Harold Mendenall	1-0-1
Lt Armour Miller	6-0-2
Lt Marvis Morrison	0-0-2
Capt. Edward Newbury	4-1-3
Lt Thomas Newell	0-1-0
Lt Frederick Nichol	1-0-2
Lt Anthony Noonan	1.166-0-0
Capt Joel Owens	5-1-1
Capt. James Pate	2-0-1
Lt. Lester Peterson	1-0-0
Lt George Plummer	0-0-2
Maj. Francis Pope	0-0-1
Lt. Col Larry Pope	0.166-0-0
Lt Thomas Prout, Jr	1-2-0
Lt William Purvis	0-2-1
Lt. Virgil R. Radcliffe	1-0-0
Capt Thomas Raphael	2-1-0
Lt Clifford Randol	1-0-1
Capt. Larry Reynolds	2.166-0-1
Lt Galen Rhoades	0-0-2
Lt Robert Riley	0-2-0
Lt George Ross	2-1-0
Capt Theodore Runyon	3-0-1
Capt George Rush	3-1-1
Lt Walter Schranz	1-0-1
Lt Elza Shahan	0-0-3
Lt Edward Shaw, Jr.	3-0-0
Lt. Earl Sprengel	2-0-0
Lt George Smith	0-1-0
Lt. Raymond Smythe	1-1-1
Lt Roger Stemen	0-1-2
Capt David Stentz	0-0-1 (1st Ftr. Grp.)
Lt Robert Stevens	1-0-0
Lt Millard Stunkard	0-1-0
Lt Irvine Styer	2-2-0
Lt John Sullivan	3-1-1
Lt. Samuel Sweet	4-0-1
Lt Ralph Tibble	1-0-1
Lt Philip Tovrea, Jr	8-1-3
Lt Eldon Vondra	1-0.5-0
Lt. Edward Ulrich	1-0-0
Lt Burton Weil	2-0-2
Lt Jerome Weinberg	1-0-1
Capt Darrell Welch	5-0-3
Lt Edward Wellner	1-1-1
Maj. John Weltman	1-0-0
Lt William Wennergren	1-2-1
Lt. Marvin Wingrove	2-1-1
Lt John Wolford	5-0-2

71st Fighter Squadron

Capt. James Advey	0-0-1
Lt Richard Ariano	2-1-1
Lt William Armstrong	1-0-0
Lt Leslie Berens, Jr	1-2-0
Lt Willard Bolton	2-0-0
Lt Joseph Boyd	1-0-0
Lt Robert Bridges	1-0-0
Lt George Brittin	1-0-0
Lt Allan Chase	2-0-0
Lt Ralph DeSutter	0-0-2
Maj John Eiland	3-0-1
Lt James Ferguson	2-0-0
Lt Rodney Fisher	5-0-1
Lt Willie Flack	1-0-0
Lt Arthur Franke	2-1-0
Lt Richard Furr	2-0.5-3
Lt Leslie Garrett	1-0-2

Lt Lee Gregg	1-0-0
Lt Emmet Gresham	1-0-0
Capt Francis Harris	0-1-0
Capt Herbert Hatch	5-1.5-3
Lt Melvin Hill	0-0-4
Lt Ralph Hisey	1-0-0
Lt Carl Hoenshell	3-0-0
Lt Robert Hoffman	1-0-0
Lt Arthur Hoodecheck	0-0-1
Lt Warren Hultman	0-0-2.5
Lt William Hunter	0-1-0
Lt Joseph Jamison	2-0-1
Lt Rudolph Janci, Jr	2-0-0
Lt George Johnson	0-1-0
Lt Raymond Jones	1-0-0
Lt William Jones	1-0-0
Lt Harold Kline	0-0-0.5
Lt Frederick Kohn	1-0-0
Lt John Launderville	1-1-2
Lt Peyton Mathis, Jr.	0-0-1
Lt Harold Melbraaton	1-0-0
Lt Joe D. Miles	3-0-0
Maj James Morehead	1-0-1
Lt Lyle Morehead	0-0-2
Lt William Morhous	0.5-0-1
Lt John Moutier, Jr.	4.5-0-2
Lt Bernard Muldoon	0.5-0-0
Lt John Neil	1-0-0
Lt Virgil Olson	1-0-1
Lt Gerald Osgood	2-0-0
Lt George Parker	1-1-1
Lt Carl Petersen	1-3-1
Lt Gerald Primm	1-0-2
Lt Vincent Rethman	0-1-0
Lt Walter Rivers	4-2-2
Lt William Royer, Jr	2-0-0
Maj Raymond Rudell	2.5-2-0
Lt Basil Rudnick	0-0-1
Lt Robert Sauer	1-0-0
Lt Edwin Schneider	1-0-0
Capt. Vern Schoenfeldt	2.5-0-0
Lt Charles Schreffler	1-2-1
Lt Meldrum Sears	7-0-0
Lt Alex Sentes	1-1-0
Lt Anthony Sgroi	1-0-0
CApt. Allan Shaw	2-0-2
Lt John Shepard	2-0-1
Capt. Charles Simpson	3-0-5
Lt Joseph Smith	0.5-0-0
Lt Thee Smith	2-0-0
Capt Robert Spitler	2-0-0
Lt Howard Standridge	0-0-1
Lt Stanley Stuber	4-0-0
Lt Glenn Tanner	4-1-1
Lt Fay Valley	0-0-1
Lt Robert Viall	1-1-0
Lt John Willey	0-1-2
Lt. Russell Williams	1-0-0
Capt Lee Wiseman	5-0-0

94th Fighter Squadron

Lt John Allen	1-0-0
Lt Richard Arrowsmith	2-0-0
Lt Herschel Baird	1-0-1
Lt Harley Barlow	4-0-3
Lt Robert Boose	1-0-1
Capt Robert Bostwick	1-0-1
Lt James Bower	3-0-1
Lt Edward Brophy	1-0-1
Lt George Burger	0-0-1
Lt Gordon Buttles	0-0-1
Lt Thomas Chambers	1-0-0
Lt John Cram	0-0-1
Lt Hugh Crandall	1-0-0
Lt Daniel Darnell, Jr.	2-0-3
Lt Fred Devenney	0-0-1
Lt James Dibble	2-1-0
Lt Charles Dollarhide	1-0-1
FO Willard Duff	1-0-0
Lt James Fairhurst	1-0-1
Lt Albert Fell	0-0-2
Lt Paul Forsythe	1-0-0
Lt Stuart Froehling	0-1-1
Lt Karl George	1-0-0
Lt Raymond Geyman	2-0-0
Lt Howard Gilliam	2-0-0
Lt Donald Greenley	1-0-1
Capt. James Hagenback	0.33-2-1
Lt Benjamin Hallock	3-1-2
Lt John Hanton	2-1-1
Capt James Harman	1-1.5-2
Lt Robert Harris	3-0-3
Lt Luther Harry	1-0-1
Lt Chester Heien	2-1-1
Lt William Hines	1-1-1
Lt Robert Holcomb	0-1.3-0.3
FO John Hopkins	1-0-1
Lt Charles Howard	1-0-1
Maj Glenn Hubbard	0.5-0-2
Lt Jack Ilfrey	5.5-0-2
Lt William Jacobsen	0-0-1
Lt Dee Johnson	0.5-0-1
Lt Donald Kienholz	6-0-1
Lt Ralph Kipper, Jr	0-0-2
Lt Kenneth Kirchhofer	1-0-0
Capt Edward LaClare	1-0-0
Lt Franklin Lathrope	5-0-1
Capt Richard Lee	5-1-0
Lt Harold Lentz	3-2-2
Lt J.C. Harrison Lentz	0-0-1
Lt William Lovell	1.5-0-1
Lt Donald Lowe	3-0-1
Lt Willis Mathews	3.5-1.3-1.3
Lt Raymond Maxfield	1-0-0
Lt Richard McWherter	2-0-0
Lt Benton Miller	2-1-0
Lt Everett Miller	5-0-0
Lt Keith Miller	2-0-0
FO Louis Mimms	0-0-1
Lt Jack Muffitt	1-0-0
Lt Robert Neale	1-0-1.5
Lt Harry Newsome	2-0-1
FO Cyril Nolen	1-1-0
Lt Harold Niles	1-0-0
Lt Stanton Opdahl	0.3-1-1
Lt John Pace	0-1-1
Lt Jack Pettus	0-0-1
Lt Cecil Quesseth	2-1-2
Lt Donald Quinn	1.5-2.5-0.5
Lt Augustus Reese	0-1-0
Capt. Clarence Rimke	1-0-0
Capt Newell Roberts	5-1-1
Lt Philip Sager	2-0-1
Lt Warren Shaper	0-1-2
Lt Raymond Shulze	1-1-1
Capt. Jerome Schweikert	2-0-0
Lt. James Smith	1-0-0
Lt Philip Smith	2-0-0
Lt Nicholas Sokol	0-0-2
Lt Donald Starbuck	1-0-1
Lt Leonard Stephan	3-0.833-1.3
Capt. George Sutcliffe	1-0.5-2
Capt. Ralph Thiessen	4-1-2
Lt Everett Umphrey	0-1-0
Lt Paul Vachon	2-0-4

APPENDIX C (CONT.)
94TH Fighter Squadron

Lt Robert Van Sice	2-0-2
Capt Robert Vrilakas	2-0-1
Capt. Otto Wellensiek	2-1-1
Lt William Whalen	1-1-0

APPENDIX D
CASUALTIES

Casualties in the 1st Fighter Group varied throughout the course of the war and can be divided into several categories, i.e., killed, wounded, prisoner of war, accident and including death from natural causes. These are indicated in the squadron lists. Taken together throughout the course of the war the Group suffered nearly 40% casualties. That is we lost 280 pilots out of a total of 720 plus some ground people wounded early on in the North Africa landing and in air raids. Losses in this context includes, in addition to the obvious, those men who were shot down but managed to escape or to avoid capture and, in time, make their way back to Allied areas-sometimes this took nearly a year of existing behind enemy lines. If anything the loss percentage is misleadingly low because of the comparatively large number of pilots taken into the Group late in the war. Some of them flew only a few missions simply because the war came to an end. On a month to month basis losses varied as much as ten fold. From the start of operations in North Africa we lost a minimum of 3 pilots in a month ranging up to a high of thirty three in August 1943. We lost no one in February 1945 largely due to the Yalta operation, bad weather and a quagmire airfield in Italy. Accidents, many of which occurred within sight of the field and were largely unexplained, accounted for nearly 10% of our total losses.

Comparisons are always difficult, however, The Army Air Forces in WW II history cites one interesting statistic which illustrates a casualty comparison late in 1943. The figure cited is the percent killed, wounded or missing in action per 1000 sorties for the period 1 Sept. 1943 through 31 Dec. 1943.

B-17	B-24	B-25	A-20	P-38
5.78	19.08	3.37	2.94	10.02

The above comparison shows rather clearly that the common assumption of greater safety and lower casualties in fighter aircraft vis a vis bombers was not necessarily true. We may have felt a bit safer watching the bombers fly a steady course through a flak barrage while we maneuvered around it but each of us lost a lot of people.

(The B-24 number seems unusually high in comparison with the B-17 figure and was not further explained. Possibly the smaller number of Liberators in the theater at that time may have skewed the statistics.)

In this appendix the pilot losses are illustrated by squadron and by yearly quarter to illustrate the loss variability. Following in the appendix the pilots and pilot losses are enumerated by squadron. The records, which date from shortly after the war, still list some of these men as missing in action which, of course, is highly unlikely. Insofar as we can determine any indicated MIA should be considered as killed in action. Early in the war we were able to learn of prisoner of war status in Germany through Red Cross and/or neutral country offices. Late in the war as Germany descended into chaos such information became increasingly more difficult to obtain thus the MIA status persisted.

Recent information obtained from the American Battle Monument Commission has enabled us to identify as missing those whose bodies were never recovered. Also included in this appendix is a listing of burials in the several American military cemeteries overseas. The missing are commemorated at these cemeteries on Tablets of the Missing.

In addition to pilot losses several men were wounded in action; others died as the result of an accident or of natural causes. These are as follows:

Sgt. Vernon Barclay, 71st Sqd. died, natural causes, England Sept. 1942

Sgt. Loren Eck, 71st Sqd. wounded by enemy sniper, Arzew landings.

Cpl. Harry Hubbard, 94th Sqd. wounded at Biskra (air raid).

Sgt. Loren Seyferth, 94th Sqd wounded at Biskra (air raid).

Sgt. George Lee, 94th Sqd. wounded at Biskra (air raid).

T/Sgt. Eli Kovacevitch, 94th Sqd. died, natural causes, Oran Nov. 1942

S/Sgt. Charles Austin, 71st Sqd. accidentally killed by rifle discharge during gun inspection.

T/Sgt. August Heath, 94th Sqd. died, natural causes Jan. 16, 1945

Pvt. Frank Miller, 27th Sqd. died natural causes February 1945.

Lt. Norbert Willig, 94th Sqd. killed in B-24 crash 23 Dec. 1944

Pilot rosters and pilot losses are recorded by squadron in the following pages.

27th Fighter Squadron

Roster of pilots June 1, 1942 to May 6, 1945

Maj John Weltman	
Capt. Joseph Glenn	WIA 2 Jan 43
Lt William Long	MIA 3 Dec. 42
Lt Roland Malmsted	
Lt Harold Mendenall	
Capt. Joel Owens	
Lt Theodore Runyon	POW 14 Jan 43
Lt Robert Stevens	
Lt Burton Weil	POW 18 Jan 43
Capt. Darrel Welch	
Lt Robert Chenowith	MIA 13 Nov. 42
Lt William Drysdale	KIA 3 Dec. 42
Lt David Everett	KIA 4 Dec. 42
Lt Elmer Hartman	
Lt William Hester	WIA 3 Dec 42
Lt Marvis Morrison	
Maj Edward Newbury	
Lt Lawrence Pace	KIA 4 Dec 42
Maj James Pate	
Lt Robert Riley	
Lt George Rose	
Lt Elza Shahan	
Lt Henry Smith	POW 2 Jan 43
Lt Roger Stemen	
Lt John Sullivan	
Lt John Wolford	KIA 19 May 43 14th F.G.
Lt Donald Field	MIA 5 Apr 43
Lt George Elkin	MIA 31 May 43
Lt Hubert Black	MIA 4 Dec. 42
Lt Marcus Linn	
Lt Donald Hilgert	KIA 5 Apr 43
Lt Herbert Gordon	MIA 19 Jan 43
Lt Eldred Loder	KIA 31 May 43
Lt John Mackay	
Lt Joseph Szarflarski	
Lt Leslie Slater	MIA 23 Mar 43
Lt Stuart Bennett	KIA 25 May 43
Lt George Bancroft	
Lt Harry Dowd	KIA 3 Mar 43
Lt Daniel Kennedy	
Lt George Rush	
Lt Robert McCord	
Lt William Martyn	KIA 3 Mar 43
Lt Virgil Radcliffe	
Lt Samuel Sweet	
Lt Robert Anderson	KIA 17 Apr 43
Lt Herbert McQuown	WIA 20 Jun 43
Lt William Purvis	
Capt Louis Burke	

Lt Warren Holden	
Lt Robert Britz	KIA 18 Jun 43
Lt Louis Bryson	KIA 6 May 43
Lt David Conn	MIA 24 Jun 43
Lt Frank McIntosh	
Lt Verne Clark	KIA 2 May 43
Lt James Meikle	
F/O Donald Deisenroth	POW 30 Aug 43
Lt Edward Shaw	POW 5 Jun 43
Lt Edward Crane	
Maj Gilbert Butler	
Lt Otto Hloucal	
Lt Robert Kuba	KIA 10 Jul 43
Lt Thurman Neilson	POW 24 Jun 43
Lt Alden Freng	KIA 25 May 43
Lt James Burke	MIA 6 May 43
Lt Jack Hall	MIA 25 May 43
Lt Ralph Tibble	
Lt Jerome Weinberg	KIA 30 Aug 43
Lt Edward Wellner	
Lt Leon Thompson	MIA 20 Jun 43
Lt Raymond Smythe	
Lt Robert Hill	
Lt Harry Warmker	KIA 30 Aug 43
Lt Walter Schranz	
F/O Millard Stunkard	
Lt Hilliard Graham	MIA 29 Aug 43
Lt John Husby	WIA 30 Aug 43
Lt William Wennergren	
Lt Lester Laird	WIA 27 Aug 43
Lt Warren Eickmann	
Lt Larry Reynolds	MIA 23 Apr 45 (2nd tour)
Lt Lester Peterson	
Lt John Hurst	MIA 23 Apr 45 (2nd tour)
Lt George Smith	
Lt Charles Brown	POW 11 Jun 43 (Retnd.)
Lt Clifford Randol	KIA 30 Aug 43
Lt Irvine Styer	
Lt Adolphus Webb	MIA 20 Aug 43
Lt Harold Post	KIA 20 Aug 43
Capt Arden Ferrin	
Capt James Alford	
Lt Eldon Vondra	KIA 1 Dec 43
Lt Merle Brown	POW 29 Mar 44
Lt Frank Gerry	POW 13 Jun 44
Capt Francis Lawson	
Capt Walter Flynn	
Lt Marcel Williams	MIA 25 Aug 43
Lt Frederick Nichol	
Maj Thomas Raphael	
Lt James Rodolff	MIA 28 Mar 44
Lt William Parsons	KIA 25 Feb 44
Lt Thomas Prout	
Lt Thomas Maloney	WIA 19 Aug 44
Lt Francis Mackle	KIA 20 Dec 43
Lt Joseph McGrath	KIA 19 Dec 43
Lt Robert Austin	
Lt Marvin Wingrove	
Lt Dwight Terry	KIA 24 Oct 43
Lt William Schnurr	
F/O Edward Morgan	
Capt Francis Maron	
Lt Augustus Einwechter	
Lt John Price	WIA 25 Feb 44
Lt Robert McIntosh	MIA 12 May 44
Lt David Fischer	MIA 21 Jan 44
Capt Armour Miller	
F/O Ronald Delaney	POW 17 May 44
Lt Harold Lindhurst	MIA 18 Mar 44
Lt Harold Lienau	
Lt Kenneth Hartwig	MIA 28 Mar 44
Lt Robert Burgoyne	
Lt Richard Cooley	KIA 12 May 44
Lt Frank Williams	WIA 18 May 44
Lt Earl Sprengel	
Lt Jim Kuykendall	KIA 28 Jul 44
Lt Duane Enyeart	
Lt Stanley Lau	
Lt John King	K 4 May 44 (accident)
Lt Richard Huber	
Lt John Kim	KIA 13 Apr 44
Lt Edward Ulrich	POW 31 May 44
Lt William Coughlin	
Lt George Plummer	
Capt Phil Tovrea	
Lt John Janson	
Lt James Joye	KIA 10 Jun 44
Lt James Lilly	POW 12 May 44
Lt Robert Rasmussen	KIA 31 May 44
Lt Harry Noone	POW 10 Jun 44
Lt Warren Perry	POW 5 Jun 44
Lt Galen Rhoades	K 21 Aug 44 (accident)
Capt August Bischoff	KIA 10 Jun 44
Lt George Eldred	KIA 13 Jul 44
Capt William Arnold	
Lt John Eliopoulis	K 29 Jun 44 (accident)
Lt William Johnson	K 3 Aug 44 (accident)
Lt Harold Lang	
Lt Leo Lemons	KIA 13 Jul 44
Lt Glenn Martin	
Lt William McClellan	WIA/POW 10 Jun 44
Lt Herbert Foor	POW 28 Jul 44
Lt William Ganley	KIA 10 Sept. 44
Lt Richard LaManna	KIA 4 Nov 44
Lt Richard Hanford	
Lt Henry Kraft	
Lt Roderick Love	
Capt David Havard	
Lt Harold Lamberson	
Capt Joe Ewers	
Lt Frank Holcombe	WIA 26 Jul 44
Lt Barton Jones	
Lt Andrew Kmonk	
Lt Harry Lewis	
Lt Anthony Noonan	
F/O Thomas Wood	KIA 13 Jul 44
Lt Col Francis Pope	
Lt James Hardin	
Capt Blair Ericson	
Lt Henry Inge	K 21 Aug 44 (accident)
Lt Edward Lowther	K 26 Oct 44 (accident)
Lt Lauren Erickson	MIA 19 Aug 44
Lt Frank Fleming	
Lt Evert Johnson	MIA 16 Nov 44
Lt Frank McHugh	MIA 1 Mar 44
Capt Richard Barr	
Lt Robert Bruce	K 3 Oct 44 (accident)
Lt Donald Collins	
Lt Robert Fittell	MIA 2 Sept 44
Lt Jack Loveless	
Capt. Edward Begley	
Capt Guy Casey, Jr	
Lt Edgar Coury	KIA 31 Mar 44
Lt Donald Lindberg	KIA 11 Oct 44
Lt Vern Nixon	KIA 2 Sept. 44
Lt John Haring	
Lt Asa Matthews	K 26 Oct 44 (accident)
Capt Robert McClement	
Lt Allen Gallier	
Lt Charles Morehouse	
Lt John Follet	KIA 21 Nov 44
Lt Erling Stensland	K 27 Nov 44 (accident)
Lt Warren Danielson	
Lt Loren Sullivan	

Lt Lavon Taylor	KIA 20 Jan 45
Lt James Thrasher	
Lt William Waddell	
Lt Joseph Wilderhold	
Lt Clem Blake	
Lt John Taylor	
Capt Chester Trout	KIA 15 Apr 45
Lt Joseph Demkovitch	
Lt John Streiff	
Lt Lewis Talansky	
Lt Maurice Terry	
Lt Kenneth Todd	
Lt Victor Wallace	
Lt Hewett Youtz	
Lt Loyd Longenbauch	
Lt Robert Carrothers	MIA 21 Jan 45
Lt Donald Kidd	
Lt William Maddox	
Lt Richard Vandeweerd	
Lt Richard Barns	
Lt James Hunt	
Lt Othel Kilpatrick	
Lt Marvin Thorson	
Lt Cary Hendrix	KIA 31 Mar 45
Lt Frred McKloskey	KIA 16 Jan 45
Lt William Lucietich	
Capt. Thomas Newell	
Lt John Price	
Lt Raymond Shank	
Lt Lee Scott	
Lt J. H. Tappan	KIA 25 Mar 45
Maj Paul Ash	MIA 18 Mar 45
Lt William Hagerman	
Lt George Lovering	
Lt Samuel Lyons	
Lt Percy Megginson, Jr	
Lt Robert Wood	
Lt George McGinnis	K 26 Jul 45 (accident)
Lt Norman Weber	
Lt Edward Smith	
Lt Francis L'Esperance	
Lt Edward Brickley	KIA 15 Apr 45
Maj. Marvin Migl	
Maj. Hugh Watson	

71st Fighter Squadron
Roster of Pilots 1 June 1942 to 6 May 1945

Maj. Raymond Rudell	
Capt Joe Miles	
Capt. John Eiland	
Lt Allen Scales	KIA 28 Dec. 42
Capt Frederick Grambo	
Lt Cortez Quickle	
Lt Bradley Pramm	
Lt Roy Russell, Jr.	
Lt Robert Sauer	
Lt Anthony Sgroi	
Lt Otto Wellensiek	
Lt William Young	MIA 2 Oct 42
Lt John Moutier	
Lt George Rush	
Lt Charles Hoey	
Lt Roger Miller	POW 12 Aug 43
Lt Francis Harris	
Lt Peyton Mathis	
Lt William Morhous	
Lt Bernard Muldoon	
Lt Vincent Rethman	
Lt Verl Schoenfelt	
Lt Meldrum Sears	
Lt Glen Stover	
Lt Raymond Parsons	
Lr Robert McCord	
Lt Robert Naismith	
Lt Luther Morefield	
Lt Joseph Smith	KIA 12 Jan 43
Lt Richard Wells	K accident (U.K.)
Lt William Newman	
Lt Leslie Garrett	
Lt Robin Hansen	
Lt James Sullivan	
Lt Frank Maglio	
Lt Wilbur Shine	
Lt Richard Ariano	
Lt Raymond Jones	
Lt Edwin Schneider	
Lt Frederic Kohn	
Lt Arthur Franke	MIA 5 Jun 43
Lt George Means	KIA 3 Apr 43
Lt Lee Gregg	
Lt Ivan Rockwell	
Lt John Willey	KIA 30 Aug 43
Lt Ernest Chapman	MIA 10 Jul 43 returned to complete missions
Lt David Diamond	KIA 10 Jul 43
Lt Garvin Peters	POW 12 Aug 43
Lt Matthew Warren	POW 28 Aug 43
Lt Walter Morrison	POW 30 Aug 43
Lt John Slack	
Capt Stanley Stuber	MIA 10 Sept 43
Lt Clarence Southgard	
Lt Russell Winegar	POW 28 Aug 43
Lt John Lion	MIA 20 Oct 43 returned
Lt Leslie Berens	
Lt Thee Smith	MIA 15 Aug 43 returned to complete missions
Lt Richard Catledge	MIA/POW 28 Aug 43 returned
Lt Charles Simpson	MIA 17 Aug 43 returned to complete missions
Lt Howard Taylor	
Lt Burdette Nygren	
Lt John Russell	
Lt Alex Michael Sentes	
Lt Robert Viall	KIA 25 Aug 43
Lt Carl Peterson	
Lt Louis Meyer	MIA 14 Jan 43
Lt Laurence Butler	KIA 1 Dec 42
Lt Ivan Salts	KIA 14 Jan 43
Maj Lee Wiseman	
Lt Harold Melbraaten	
Lt Alden Landers	KIA 22 Jan 43
Lt William Dunn	MIA 22 Jan 43
Lt Henry Beeson	
Lt George Britton	
Lt John Staiger	
Lt Walter Rivers	
Lt William Royer	
Lt David Wilson	KIA 25 Feb 43
Lt Willard Bolton	
Lt Charles McCann	
Lt Edward Helm	
Lt Lindo Pompili	
Lt Allen Chase	
Lt James Connor	KIA 3 Oct 43
Lt Charles Schreffler	
Lt Basil Rudnick	KIA 28 Aug 43
Lt Ralph Turrentine	KIA 30 Aug 43
Lt Rodney Fisher	
Lt Donald Craft	
Lt Frederick Messmore	MIA 9 Sept 43
Lt Joseph Boyd	MIA 4 Sept 43 returned to complete missions
Lt Richard Shearer	
Lt Col Horace Hanes	MIA 7 Jan 44 evaded

Lt Hobart Vincent	KIA 3 Oct 43
Lt Russell Williams	K 29 Nov 43 (accident)
Lt Ralph Parrish	KIA 3 Oct 43
Lt Lyle Morehead	KIA 17 May 44
Lt. Jesse Spaulding	MIA 21 Oct 43
F/O Lee Rowe	MIA 7 Jan 44
Lt. James McClure	KIA 15 Nov 43
Lt John Shepard	MIA 16 Jun 44
Capt. Robert Spitler	
Lt Gerald Primm	
Lt Edward Tanner	POW 1 Dec. 43
Lt Howard Standridge	
Lt John Eastman	
Lt Emmet Gresham	MIA 25 Feb 44 evaded & returned
Lt Junior Downing	MIA 20 Oct 43 evaded & returned
Lt James Ferguson	
Lt Warren Hultman	
Lt Paul Binsley	MIA 11 Nov 43 (local)
Lt Richard Furr	
Capt James Advey	
Lt Horace Corbett	KIA 7 Jan 44
Lt Reed Owens	POW 7 Jan 44
Lt Gerald Osgood	
Lt George Parker	POW 1 Dec 43
Lt Thomas Collier	KIA 10 Feb 44
Lt Fay Valley	KIA 6 Apr 44
Lt Albert Smith	MIA 10 Jun 44
Lt Robert Bridges	
Lt Larsson	K 3 Apr 44 (accident)
Lt Raymond Dart	
Lt William Newberry	
Lt Col James Morehead	
Lt Carl Hoenshell	KIA 10 Jun 44
Lt Charles Simpson	
Lt Donald Kuske	KIA 13 Jun 44
Lt George Johnson	MIA 10 Jun 44
Lt Glenn Tanner	
Lt Herbert Hatch	
Lt Oscar Johnson	
Lt Joseph Jackson	KIA 10 Jun 44
Lt Joseph Morrison	
Lt George O'Connor	
Lt Franklin Pate	KIA 17 May 44
Lt Joseph Jamison	
Lt John Neil	
Lt Ralph DeSutter	
Lt Ralph Hisey	POW 10 Jun 44
Lt Allen Shaw	
Lt John Launderville	
Lt Robert Berg	K 28 May 44 (accident)
Lt Melvin Hill	
Lt Daniel Dixon	
Lt Vernon Baker	KIA 10 Jun 44
Lt Rudoph Janci	POW 10 Jun 44
Lt Willie Flack	KIA 10 Jun 44
Lt William Potter	KIA 10 Jun 44
Lt Robert Yaryan	K 30 Jun 44 (accident)
Lt Harold Kline	MIA 18 Aug 44
Lt Paul Kimball, Jr	Died of injuries 14 Aug 44
Lt William Armstrong	
Lt Myron Hutchins	
Lt Arthur Hoodecheck, Jr	MIA 2 Sept 44 evaded & returned.
Lt Earnest Middleton	
Lt Gerald Dodson	
Lt Robert Longworth	
Lt John Livingston,Jr	
Lt Virgil Olson	KIA 20 Oct 44
Lt William Jones	
Lt Jesse Dorris	KIA 31 Aug 44
Maj. Joseph Elliott	KIA 15 Apr 45

Lt Ronald Brewer	
Lt Richard Decker	MIA 23 Oct 44
Lt Paul Lambert	
Lt Roger Amadon	
Lt Leo Brindley	MIA 12 Sept 44
Lt Jack Anderson	
Capt Clarence Knapp	KIA 23 Apr 45
Lt Roger Lang	
Lt Harrison Fisher	
Lt Warren Campbell	
Lt Carlton Burns	
Lt Carl Alexander	
Lt Col Frank Jones	
Lt Robert Hoffman	
Lt Sidney Howell	
Lt Raymond Hurley	
Lt Felix Usis	
Lt Lowell Twedt	KIA 20 Oct 44
Lt William Wisner	KIA 20 Oct 44
Lt Paul Summer	
Lt William Ward	
Lt Thomas Dunne	
Lt William Ferguson, Jr	
Lt John Broderick	KIA 16 Jan 45
Lt Herbert Jarvis	
Lt Stuart Clark	
Lt Roy Marsh	
Lt Angelo Scalia	
Lt James Nash	
Lt Norman Crawford	MIA 31 Mar 45
Lt Rolla Griffith	
Lt Ronald Brown	
Lt Edward Steffani	MIA 16 Nov 44
Lt Edmund Roy	
Lt Raymond Douglas	
Lt James Graham	
Lt Carl Hagen	
Lt John Hite	
Lt Adrian Hulverson	
Lt Gerald Isbell	KIA 5 Apr 45
Lt Royden Kirkpatrick	
Lt Robert Lamphere	
Lt James Manley	
Lt Robert McIntyre	
Lt John Hollon	
Lt Steven Dostal	MIA 18 Mar 45 (evaded, returned
Lt David Grace	
Lt Marvin Greene	
Lt Vernon Henderson	K 23 Dec. 44 (accident)
Lt James Hulslander	
Lt Robert Huntington	
Lt Joseph Kepford	
Lt Warren Kohlman, Jr	
Lt Pietrino Maimone	
Lt John Manning	
Lt Richard Moore	
F/O Charles Miskulin	
Capt Henry Chateau	
Lt Roger Cook	
Lt Charles Kemen, Jr	
Lt George Coles, Jr	
Lt Wallace Evans	
Lt Donald Griffis	
Lt Andrew Kahanek	
Lt George McGrew	KIA 15 Apr 45
Lt Robert Phillips	
Lt James Hoagland	
Lt Lloyd Krause	
Lt Norman Moore	
Lt Cary Barrett	
Lt Kenneth Howe	

71st Fighter Squadron (cont.)

Lt Curry
Lt Robert Gosslee
Lt Col Robert Baldwin
Lt Lloyd Longenbaugh
Lt Charles McCrary
Lt William Crisp
Lt Robert Dean

94 th Fighter Squadron
Roster Of Pilots 1 June 1942 - 6 May, 1945
(includes casualties from Dec. 1941)

Maj. Ralph S. Garman	
Maj. Glenn E. Hubbard	
Lt Foy Robertson	KIA 21 Dec. 41
Lt William Northern	KIA 21 Dec. 41
Lt James Lauck	KIA 12 Mar. 42
Lt Lawrence Scholtzhauer	KIA 29 Mar. 42
Lt. Walter Lichtenberger	KIA 30 Apr. 42
Capt. James Harman	KIA 17 Apr. 43
Capt. Clarence R. Rimke	KIA 22 Feb. 43
Capt. Newell O. Roberts	
Capt. Robert E. Williams	MIA 3 Dec.42
Lt George W. Sutcliffe	WIA 18 Dec. 42 completed missions
Lt Charles W. Oakley	K 19 Aug 42 accident UK
Lt William Pennington	K 11 Oct. 42 accident UK
Lt Lewis C. Murdock	
Lt Robert D. Neale	
Lt Wesley M. Pringle	
Lt Everett F. Umphrey	KIA 12 DEc. 42
Capt. Jack Ilfrey	
Lt Earle W. Hille	
Lt Richard W. McWherter	K 15 Jan 43 accident
Lt Norman L. Widen	MIA 18 Dec. 42 POW
Lt Donald O. Starbuck	
Lt J.C. Harrison Lentz	MIA 7 Jan 43 returned
Lt Jack S. Longseth	KIA 3 Dec. 42
Lt Clifford D. Mohlzahn	
Lt James K. Smith	
Lt Clark O. Jennings	MIA 18 Dec. 42 returned
Lt Edward S.E. Newbury	
Lt Robert H. Wilson	
Lt William A. Lovell	KIA 11 Jan 43
Lt Victor J. Giles	MIA 18 Dec. 42
Lt Paul J. Forsythe	
Lt Milton A. Sahl	KIA 26 Dec. 42
Lt Richard B. Jones	MIA 26 Dec. 42
Lt Harold Lentz	MIA 25 Mar. 43 returned, completed missions
Lt Willis E. Mathews	
Lt Thomas E. Chambers	MIA 22 Feb 43 POW
Lt William H. Wilson	MIA 11 Jan 43
Lt Carl E. Peterson	
Lt Edwin R. Crane	
Lt Benton Miller	MIA 10 Mar 43 returned, completed missions
Lt Donald M. Quinn	
Lt Robert I. Boose	
Lt Jack R. Brown	
Lt Archie G. Jackson	
Lt William W. Whalen	
Lt Albert W. Fell	
Lt Leonard P. Stephen	
Lt Dee A. Johnson	KIA 11 Jul 43
Lt Max S. Rayburn	WIA 25 May 43
Lt Edward G. Brophy	
Lt Robert G. Holcombe	KIA 15 Jul 43
Capt. James J. Hagenback	
Lt George B. Burger	KIA 10 May 43
Lt Donald C. Lowe	KIA 22 May 43
Lt Harry C. Newsome	
Lt Augustus F. Reese	KIA 14 May 43
Capt Otto F. Wellensiek	
Lt Stuart C. Froehling	
Lt James W. Cronin	POW 20 Jun 43
Lt Philip A. Sager	POW 20 Aug 43 escaped, returned
Lt James P. Dibble	KIA 9 Sept 43
Lt Stanton N. Opdahl	
Lt Edward O. Esser	POW 20 Jun 43
Lt Emory L. Cogswell	
Lt Glen R. Stover	
Lt Gerry H. Hammond	
Lt Earl E. Holms	
Lt Robert H. Anderson	
Lt Howard A. Gilliam	KIA 9 Jul 43
Lt Nicholas Sokol	
Lt Harold P. Niles	
Lt James E. Bower	
Capt. Gilbert Pritchard	
Lt John D. Pace	
Capt. Ralph L. Kipper	
Capt. William L. Jacobsen	
Lt John A. Hay	MIA 24 Jun 43
Lt William W. Greishaber	KIA 10 Aug 43
Capt John T. Hanton	
Lt John H. Wilson	POW 10 Aug 43
Lt Harold A. Herr	MIA 5 Aug 43 escaped, returned
Lt John G. Cram	MIA 30 Aug 43
Lt Robert W. Boggess	KIA 15 Jul 43
Lt Marion E. Phillips	MIA 11 Sept 43 POW
Capt Richard J. Lee	
Lt Ralph E. Peck	MIA 30 Aug 43
Lt Glenn E. Terry	KIA 10 Aug 43
Capt Robert Vrilakas	
Lt Robert V. Van Sice	
Lt Roane T. Sias	MIA 20Aug 43 escaped, returned
Capt. Ralph A. Thiessen	
Capt. Raymond L. Schulze	
Lt David O. Parlett	MIA 29 Aug43 escaped, returned
Lt Jack D. Pettus	
Capt. Donald D. Kienholz	
F/O Robert E. Allan	KIA 20 Aug 43
Lt James Griffis	KIA 7 Jan 44
Lt Charles A. Woodard	MIA 30 Aug 43
Lt Harold D. Rigney	MIA 30 Aug 43escaped, returned
Lt. Warren L. Schaper	MIA/POW 17 May 43
Lt Walter D. Pittman	
Capt Jerome G. Schweikert	
F/O Willard R. Duff	MIA 17 May 43
Capt. Charles L. Hoffman	
Capt. Charles W. Howard	
Lt Stanley W. Wojcik	MIA 9 Sept 43 POW transferred
Lt John McEachern	
Lt Jack P. Muffitt	KIA 3 Jan 44
F/O Cyril L. Nolen	MIA 20 Apr. 44 POW
Capt. Daniel Darnell	
Lt Robert D. Harris	
Lt Fred Devenny	KIA Jan 7 44
Capt Harley T. Barlow	
F/O Louie Mimms	
Lt William E. Hines	
Lt Lipowicz	KIA 20 Dec. 43
F/O Charles H. Brown	KIA 20 Dec. 43
Lt Harmer	KIA 20 Dec. 43
Lt Raymond J . Meredith	MIA 7 Jan 44 escaped,

	returned
Lt Kenneth M. Kirchhofer	
Lt Kastner	transferred
Capt. Cecil H Quesseth	
Capt Robert L. Laurent	
Lt Raymond J. Geyman	MIA 10 Jun 44 POW
Lt Paul Vachon	
Capt Franklin C. Lathrope	WIA/MIA 10 May 44,, evaded, returned
Lt Herschel W. J. Baird	
Capt Benjamin Hallock	WIA 18 Jun 44
Lt John H. Allen	
Lt Warren G. Campbell	
Lt Luther M. Harry	KIA 17 Apr 44
Lt John A. Hopkins	WIA Jun 44
Maj Edward F. LaClare	
Lt Harold V. Duggleby	KIA 15 Jun 44
Lt Don H. Greenley	
Lt Karl O. George	
Lt Donald Hammerstrom	KIA 17 May 44
Lt Philip W. Smith	
Lt Raymond M. Maxfield	MIA 5 Jun 44 evaded, returned
Capt Everett Miller	
Lt Keith R. Miller	
Lt Richard Arrowsmith	WIA/MIA 18 Aug 44 evaded, returned
Lt James A. Fairhurst	W Dec. 44 accident
Lt William M. Bechtold	
Lt Samuel Barnes	KIA 16 Jun 44
Lt George Loughmiller	KIA 16 Jun 44
Lt Marlyn Kobiske	K 19 Jun 44 Accident
Lt Charles L. Dollarhide	
Lt Hugh G. Crandall	KIA 15 Jun 44
Lt Thomas Vitale	KIA 16 Jun 44
Lt Chester H. Heien	
Lt Royal M. Nyby	
Lt Carrol W. Feather	
Lt John J. Hawkins	
Lt Russell E. Jensen	MIA 11 Nov 44
Lt Anthony W. Kainrad	K 23 Dec 44 accident
Lt Leo F. Karl	
Lt R. G. Gillen	KIA 31 Mar 45
Lt Lewis F. Kirby	
Capt. Robert C. Bostwick	
Lt Gordon L. Buttles	
Lt Robert L. Cake	
Lt Eugene M Chambers	K 24 Jul 44 accident
Lt Howard H. Dale	MIA 11 Oct 44 POW
Lt Jack Hughes	MIA 16 Jul 44
Capt John D. Mullins	
Lt Walter Pregmon	
Lt Howard G. Micheli	MIA 13 Jul 44
Lt David H. Raner	
Lt William G. Wagnecz	
Lt Richard G. Sykes	MIA 6 Dec. 44 returned
Lt Robert D. Taylor, Jr	MIA 16 Aug 44
Lt Walter O. St John	MIA 20 Aug 44
F/O Louis L. Driggs	MIA 6 Dec 44 evaded, returned
Lt Walter M. Gonring	WIA 19 Aug 44
Lt Chester Kozerski	
Lt Allen F. Long	
Lt Frederick Aubele	
Lt Robert V. Clifford	
Lt Arvid P. Evenson	K 7 Sept 44 accident
Lt Robert A. Hester	
Lt Lyle H. Baker	
Lt William E. Dougherty	
Lt Leroy W. Eberle	
Lt Jack R. Fehrenback	
Lt Robert E. Leen	WIA 21 Nov. 44
Capt. Everett Lindley	WIA/ 15 Apr 45 POW
Lt. Col Francis H. Harris	
Lt Robert V. Jurgensen	K. 25 Oct. 44 accident
Lt Willliam Ford	K. 23 Nov. 44 accident
Lt Edwin Rosenthal	
Lt Guy J. Thomas	KIA 26 Nov 44
Lt Kenneth C. Vradenburg	
Capt. Arnold R. Megenity	
Lt Victor Bennett	
Lt Clifford R. Crowl	
Lt Jack K. Faxon	
Lt Charles J. Morris	
Lt Roger R. Jones	
F/O Maurice H.J. Sevin	
Lt Ben D. Balthis, Jr.	
Lt Harry V. Cheshire, Jr.	
Lt James B. Jackson	
Lt Harvey W. McDaniel	
Lt Robert F. Taylor	
Lt Edwar B. Faulkner	
Lt Lyle F. Finley	
Lt James F. Franey	
Lt Roy M. Mortenson	
Lt Earle N. Parker	
Lt Vincent A. Skelpsa	
Lt Gordon E. Franklin	
Lt Harry A. Eberhardt III	
F/O Thomas R. Graffam	KIA 21 Jan 45
Lt John D. Felsinger	KIA 9 Mar 45
Lt Norman D. Olson	
Lt Weldon W. Schorg	WIA date indef.
Lt Robert C. Dickmann	
Lt Philip W. Fischer	
Lt Ramond M. Cleary	
F/O Carl Hilashi	
Lt James Hutchinson, Jr	MIA 21 Jan 45
Lt Rolla E. Griffith	
Lt Jack Grooms	
Lt James L. Hile	
Lt Edgar W. Hoge	
Lt Elsworth M. King	
Lt William R. Greshong	
Lt Joseph D. Kirby	
Lt Cyril E. Williams	
Lt Earl B. Hill	
Lt Arthur M. Johnson	
Lt Jack M. King	
Lt Arthur B. Sweeter	
Lt Robert P. Weaver	
Lt Ralph M. Green	
Lt Marvin T. Johnson	
Lt Shaw	
Lt Col. Harry L. Culberson	

American Battle Monuments Commission

First Fighter Group casualties buried or commemorated in military cemeteries overseas have been identified by the American Battle Monuments Commission. The cemeteries and those buried or remembered there are given below. The designation MIA indicates that the body was never recovered and the name is inscribed on the Roll of the Missing at that particular cemetery.

Ardennes American Cemetery

Name & Rank		Squadron
Hendrix, Kary L.,	Lt.	27th Ftr. Sqd.
Rasmussen, Robert P.,	Lt	27th Ftr. Sdq.
Baker, Vernon C.	Lt.	71st Ftr.Sqd.
Potter, William,	Lt.	71st Ftr. Sqd.

Barnes, Samuel D.,	F.O.	94th Ftr. Sqd.
Felsinger, John D.,	Lt.	94th Ftr. Sqd.

Cambridge American Cemetery

Name & Rank	Squadron
Wells Richard H.,Capt.	71st Ftr Sqd.

Florence American Cemetery

Name & Rank		Squadron	
Cooley, Richard A.	Lt	27th Ftr. Sqd	
Follett, John R.	Lt	27th Ftr. Sqd	MIA
Kuykendall, Jim L	Lt	27th Ftr. Sqd	MIA
Lindberg, Donald	Lt	27th Ftr. Sqd	MIA
McGrath, Joseph P.	Lt	27th Ftr. Sqd.	
Parsons, William G.	Lt	27th Ftr. Sqd.	MIA
Terry, Dwight E.	Lt	27th Ftr. Sqd.	
Vondra, Eldon E.	Lt	27th Ftr. Sqd.	MIA
Collier, Thomas	Lt	71st Ftr. Sqd.	MIA
Dorris, Jesse R.	Lt	71st Ftr. Sqn.	
Hoenshell, Carl C.	Lt.	71st Ftr. Sqd.	MIA
Knapp, Clarence I.	Capt.	71st Ftr. Sqd.	
Morehead, Lyle E.	Lt.	71st Ftr. Sqd.	
Pate, Franklin E.	Lt.	71st Ftr. Sqd.	
Smith, Albert F.	Lt.	71st Ftr. Sqd.	MIA
Steffani, Edward	Lt.	71st Ftr. Sqd.	MIA
Twedt, Lowell S.	Lt.	71st Ftr. Sqd.	MIA
Valley, Fay R.	Lt.	71st Ftr. Sqd.	MIA
Wisner, William O	Lt.	71st Ftr. Sqd.	MIA
Graffam, Thomas R.	F/O	94th Ftr. Sqd.	
Hammerstrom, Donald A.	Lt.	94th Ftr. Sqd.	MIA
Muffit, Jack P.	Lt.	94th Ftr. Sqd.	

Lorraine American Cemetery

Name & Rank		Squadron
Brickley, Edward G.	Lt	27th Ftr. Sqd.
Ganley, William J.	Lt	27th Ftr. Sqd.
Trout, Chester E.	Capt.	27th Ftr. Sqd.
Elliott, Joseph R.	Maj.	71st Ftr. Sqd.
Loughmiller, George	Lt.	94th Ftr. Sqd.

North African American Cemetery

Name & Rank		Squadron	
Bennett, Stuart M.	Lt.	27th Ftr. Sqd.	MIA
Black, Hubert R.	Lt	27th Ftr. Sqd.	MIA
Britz, Robert C	Lt.	27th Ftr. Sqd.	MIA
Clark, Vern R.	Lt	27th Ftr. Sqd.	
Conn, David M	Lt	27th Ftr. Sqd.	MIA
Dowd, Harry E.	Lt.	27th Ftr. Sqd.	
Drysdale, William	Lt	27th Ftr. Sqd	
Everett, David R.	Lt	27th Ftr. Sqd.	
Field, Donald E.	Lt	27th Ftr. Sqd.	MIA
Freng, Alden E.	Lt.	27th Ftr. Sqd.	MIA
Hall, Jack C.	Lt.	27th Ftr. Sqd.	MIA
Hilgert, Donald B.	Lt.	27th Ftr. Sqd.	MIA
Loder, Eldred E.	Lt.	27th Ftr. Sqd.	MIA
Long, William K.	Lt.	27th Ftr. Sqd.	MIA
Martyn, William H.	Lt.	27th Ftr. Sqd.	
Post, Harold F.	Lt.	27th Ftr. Sqd.	MIA
Randol, Clifford E.	Lt.	27th Ftr. Sqd.	MIA
Slater, Leslie V.	Lt.	27th Ftr. Sqd.	MIA
Thompson, Leon E.	Lt.	27th Ftr. Sqd.	MIA
Weinberg, Jerome N.	Lt.	27th Ftr. Sqd.	MIA
Binsley Paul W.	F/O	71st Ftr. Sqd.	MIA
Butler, Lawrence A.	Lt.	71st Ftr. Sqd.	
Franke, Arthur C.	Lt.	71st Ftr. Sqd.	MIA
Landers, Alden L.	Lt.	71st Ftr. Sqd.	MIA
Means, George P	Lt.	71st Ftr. Sqd.	
Meyer, Louis F.	Lt.	71st Ftr. Sqd.	MIA
Parrish, Ralph F	Lt.	71st Ftr. Sqd.	MIA
Rudnick, Basil S.	Lt.	71st Ftr. Sqd.	MIA
Salts, Ivan W.	Lt.	71st Ftr. Sqd.	MIA
Smith, Joseph R.	Lt.	71st Ftr. Sqd.	MIA
Willey, John S.	Lt.	71st Ftr. Sqd.	MIA
Burger, George B.	Lt.	94th Ftr. Sqd.	MIA
Giles, Victor G.	Lt.	94th Ftr. Sqd.	MIA
Harman, James	Maj.	94th Ftr. Sqd.	MIA
Longseth, Jack S.	Lt.	94th Ftr. Sqd.	MIA
Lovell, William A	Lt.	94th Ftr. Sqd.	MIA
Lowe, Donald C.	Lt	94th Ftr. Sqd.	MIA
Peck, Ralph E.	Lt	94th Ftr. Sqd.	MIA
Phillips, Marion E.	Lt.	94th Ftr. Sqd.	MIA
Rimke, Clarence	Capt.	94th Ftr. Sqd.	MIA
Sahl, Milton A	Lt.	94th Ftr. Sqd.	MIA
Umphrey, Everett F.	Lt	94th Ftr. Sqd.	MIA
Wilson, William H	Lt	94th Ftr. Sqd.	MIA
Stentz, David S.	Capt	HQ 1 Ftr. Grp.	MIA

Rhone American Cemetery

Name & Rank		Squadron	
Crandall, Hugh B	Lt.	94th Ftr. Sqd.	
Hay, John A.	Lt.	94th Ftr. Sqd.	MIA
St John, Walter O.	Lt.	94th Ftr. Sqd.	MIA

Sicily-Rome Cemetery

Name & Rank		Squadron	
Bruce, Robert F.	Lt.	27th Ftr. Sqd.	
Bryson, Louis M.	Lt.	27th Ftr. Sqd	MIA
Burke, James W.	Lt.	27th Ftr. Sqd.	MIA
Fischer, David J	Lt.	27th Ftr. Sqd.	MIA
Graham, Hilliard	Lt.	27th Ftr. Sqd.	MIA
Kuba, Robert J.	Lt	27th Ftr. Sqd.	MIA
LaManna, Richard F.	Lt.	27th Ftr. Sqd.	
Lowther, Edward K	Lt.	27th Ftr. Sqd.	
Matthews, Asa H.	Lt.	27th Ftr. Sqd.	
McIntosh, Robert L.	Lt.	27th Ftr. Sqd.	MIA
Rhoades, Galen J.	Lt.	27th Ftr. Sqd.	
Taylor, Lavon O.	Lt.	27th Ftr. Sqd.	
Warmker, Harry D.	Lt.	27th Ftr. Sqd.	MIA
Webb, Adolphus M.	Lt.	27th Ftr. Sqd.	MIA
Conner, James G.	Lt	71st Ftr. Sqd.	MIA
Corbett, Horace S.	Lt.	71st Ftr. Sqd.	
Diamond, David M.	Lt	71st Ftr. Sqd.	
McClure, James A.	Lt.	71st Ftr. Sqd.	
Spaulding, Jesse E.	F/O	71st Ftr. Sqd.	MIA
Turrentine, Ralph F.	Lt.	71st Ftr. Sqd.	MIA
Viall, Robert E.	Lt.	71st Ftr. Sqd.	MIA
Vincent, Hobart S.	Lt.	71st Ftr. Sqd.	MIA
Williams, Russell E.	Lt.	71st Ftr. Sqd.	
Yaryan, Robert H.	Lt.	71st Ftr. Sqd.	
Allan, Robert F.	F/O	94th Ftr. Sqd.	
Cram, John G.	Lt.	94th Ftr. Sqd.	MIA
Dornon, Ford C.	Cpl.	94th Ftr. Sqd.	
Gillen, R. G.	Lt.	94th Ftr. Sqd.	MIA
Gilliam, Howard A.	Lt.	94th Ftr. Sqd.	MIA
Grieshaber, William W.	Lt.	94th Ftr. Sqd.	
Holcombe, Robert J.	Lt.	94th Ftr. Sqd.	
Jensen, Russell E.	Lt.	94th Ftr. Sqd.	MIA
Johnson, Dee A.	Lt.	94th Ftr. Sqd.	MIA
Jurgensen, Robert V.	Lt	94th Ftr. Sqd.	
Kobiske, Marlyn E.	Lt.	94th Ftr. Sqd.	
Michele, Howard G.	Lt.	94th Ftr. Sqd.	MIA
Willig, Norbert J.	Lt.	94th Ftr. Sqd.	

Cemetery locations are as follows:
Ardennes: twelve miles southwest of Liege, Belgium .
Cambridge: three miles west of Cambridge, England

Florence: seven and one-half miles south of Florence, Italy
Lorraine: three-quarters of a mile north of St. Avold, France (28 miles east of Metz)
North Africa: ten miles from Tunis near the site of ancient Carthage.
Rhone: Draguignan, France
Sicily-Rome: immediately east of Anzio, Italy

APPENDIX E

GROUP AND SQUADRON COMMANDERS

1st Fighter Group Commanders

Lt Col Robert Israel	1May 41
Maj. John Zahn	1 May 42
Col. John Stone	9 Jul 42
Col Ralph Garman	7 Dec 42-1 Sept 43
Col. Robert Richard	19 Sept 43
Col Arthur Agan	15 Nov 44
Lt Col Milton Ashkins	31 Mar 45
Lt Col Charles Thaxton	11 Apr 45
Col Milton Ashkins	28 Apr 45

Deputy Group Commanders

Lt Col Wilbur Stratton
Lt Col H.F. Van Leuven
Lt Col Milton Ashkins
Lt Col Charles Thaxton
Lt Col Charles Slocumb

27th Squadron Commanding Officers

Major William Covington	15 Jan. 1942
Major John Weltman	15 Jan 1942 - 5 Dec.1942
Capt. Joseph Glenn	5 Dec. 1942 - 3 Jan. 1943
Capt Joel Owens	3 Jan. 1943 - 26 Mar. 1943
Capt James Pate	27 Mar. 1943 - 1 Sept. 1943
Major Edward Newbury	1 Sept. 1943 - 27 Sept. 1943
Major Gilbert Butler	27 Sept. 1943 - 11 Apr. 1944
Major Thomas Raphael	11 Apr. 1944 - 20 July 1944
Lt Col Francis Pope	20 July 1944 - May 1945

71st Squadron Commanding Officers

Capt. Frederick Rambo	1 Feb. 1942
Major Raymond Rudell	1 Feb. 1942 - 9 Mar. 1943
Major John Eiland	9 Mar. 1943 - 1 June 1943
Major Joe Miles	1 June 1943 - 27 Sept. 1943
Major Horace Hanes	27 Sept. 1943 - 9 Dec. 1943
Major Lee Wiseman	9 Dec. 1943 - Sept. 1944
Lt Col Frank Jones	Sept. 1944 - May 1945

94th Squadron Commanding Officers

1st Lt Ralph Garman	9 Dec. 1941 - 31Jan. 1942
1st Lt Glenn Hubbard	31 Jan 1941 - 6 Mar. 1942
Capt. James Harman	6 Mar. 1943 - 17 Apr. 1943
Capt George Sutcliffe	17 Apr. 1943 -1 July 1943
Maj. Otto Wellensiek	1 July 1943 - 27 Sept. 1943
Capt. James Hagenback	27. Sept. 1943 -24 Jan.1944
Capt. William Jacobsen	24 Jan 1944 - 28 Apr. 1944
Capt Edward LaClare	28 Apr. 1944 - 20 July 1944
Lt. Col. Francis Harris	20 July 1944 - 21 May 1944

APPENDIX F
STAFF OFFICERS

1st Fighter Group

Lt Col Ralph Garman	Executive Officer
Lt Walter Roney	"
Lt Wilson	Adjutant
Maj Charles Bornhorst	"
Maj Lawrence Peterson	"
Maj Arthur Miller	"
CWO Ellis Burns	"
Maj Clyde Gabler	Intelligence
Maj A. B. Sherry	"
Maj William Richardson	
Capt James Carton	"
Capt. C.J. Chamales	"
Capt Herbert Goldman	"
Capt George Townsend	"
Capt Albert Hall	"
Lt John Hardt	"
Capt Donald Long	"
Capt Henry Neck	"
Maj Charles Beazley	"
Capt John Calderbank	Operations
Lt Col Ralph Garman	"
Maj Walter Roney	"
Lt Col Horace Hanes	"
Lt Col Burton McKenzie	"
Lt Col William Stratton	"
Lt Col James Morehead	"
Lt Col Lee Wiseman	"
Maj Jerome Schweikert	"
Maj Hugh Watson	"
Capt David Stentz	"
Capt George Rush	"
Capt James Pate	"
Capt James Alford	"
Capt Francis Maron	"
Capt James Advey	"
Lt Gerald Dodson	"
Lt Gilbert Norris	"
Maj Harry Culbertson	"
Maj Robert Baldwin	"
Capt Elfege Austad	Engineering
Capt Frederick Carr	"
Capt Herman Smith	Weather
Capt Robert Bixler	"
Capt Edward McManmon	Dental Surgeon
Capt Joseph Ruebel	Surgeon
Maj. Nelson	"
Maj Emil Sposato	"
Maj Sherman Masler	"
Capt Delbert Kelly	Chaplain
Elbert Morford	"
Jack Bodart	"

Maj Paul Fitzpatrick	Communications

27th Squadron Staff Officers

Maj. Andrew Johnson	Executive officer
Capt Andrew Johnson	Adjutant
Capt Edward Egelko	"
Capt William Cody	Intelligence POW 13 Apr43
Maj William Richardson	"
Lt Nathan Pelcovits	"
Capt Samuel Harrison	"
Capt Richard Peterson	Engineering
Capt Thomas Prout	"
Lt Charles Allen	Communications
Lt William Handelman	"
Lt Charles Pappas	Ordnance
Lt Eugene Goebel	Armament
Lt Charles Duke	Supply
Lt John Pegram	"
Lt Donald Colbourne	"
Capt Delbert Kelly	Surgeon
Capt Archibald Adams	"
Capt Joseph Glenn	Operations
Capt Edward Newberry	"
Capt Gilbert Butler	"
Capt Thomas Raphael	"
Capt David Havard	"
Capt Robert McClement	"

71st Squadron Staff Officers

Capt Robert Watts	Executive Officer
Maj. George Bisbee	"
Lt James Moore	Adjutant
Capt Lawrence Peterson	"
Capt Robert Watts	"
Capt John Cunningham	"
Capt Robert Snyder	Intelligence
Capt James Carton	" "
Capt William Richardson	" "
Capt Edgar Christian	" "
Lt Elfege Austad	Engineering
Lt Pershing McAllister	" "
CWO Carey Langston	"
Capt Frederick Carr	"
Capt Victor Egasti	"
Lt David Thomas	Communications
Lt Melvin Parsons	"
Lt Edward May	Ordnance
Capt Edward Stratemeier	Surgeon
Capt Joe Miles	Operations
Maj. Joseph Elliot	"

94th Squadron Staff Officers

Maj. Jerome Belleau	Executive Officer
Capt Arthur Miller	Adjutant
Capt John Armstrong	"
Lt Walter Tattersoll	"
Capt Ronald McEwen	Intelligence
Capt Thomas Ervin	"
Maj. Charles Beazley	Intelligence
Capt Frederick Carr	Engineering
Lt Welton Avery	"
Lt William Wagnecz	"
Capt Bruce Brown	"
Lt Lawrence Trout	Communications
Lt John Hotuiz	Ordnance
Lt Norbert Willig	Armament
Capt. - Borup	Supply
Capt Horace Wood	"
Capt Kirk Garretson	Surgeon
Capt Edward Walsh	Surgeon
Lt Jack Grooms	Personal Equipment
Capt Jack Harmon	Operations
Capt James Hagenback	"
Capt Otto Wellensiek	"
Capt William Jacobsen	"
Maj Edward LaClare	"
Capt Robert Bostwick	"

APPENDIX G

CAMPAIGNS

Antisubmarine, American Theater
Egypt-Libya
Air Offensive, Europe
Algeria, French Morocco
Tunisia, Sicily
Naples, Foggia, Anzio
Rome, Arno
Normandy
Northern France
North Apennines
Rhineland
Central Europe
Po Valley
Air Combat, EAME (Europe, Africa, Mediterranean) Theater

BIBLIOGRAPHY

* **Fighters Over Tunisia**, Christopher Shores, Hans Ring, William Hess; Neville Spearman Ltd. London
* **The Army Air Forces in World War II** Univ. of Chicago Press
* 1st Fighter Group Histories; monthly record of events of the 27th, 71st and 94th Fighter Squadrons; USAF Historical Research Center, Maxwell AFB, Alabama
* **History of the First Fighter Group Vols. 1 & 2** James Alford, unpublished.
* 1st Fighter Group and Squadron mission reports on file at National Archives, Washington DC (not microfilmed)
* **Air War**, Edward Jablonski; Doubleday & Co. Garden City, NY
* **Impact, The Army Air Forces Confidential Picture History of World War II** James Parton & Co. NY,NY
* **Fifteenth Air Force Story,** Historical Aviation Album
* **Luftwaffe**, Williamson Murray; Nautical Aviation Publ. Co. Baltimore, Md.
* **The Luftwaffe War Diaries**, Cajus Becker, Doubleday & Co. Garden City, NY
* **The Second D-Day**, Jacques Robichon; Walker & Co. NY, NY
* **Happy Jack's Go Buggy,** Jack Ilfrey; Exposition Press, Hicksville, NY
* Kingscliffe Remembered, Newsletter of the 20th Ftr. Grp.
* **What Did You Do In WW II Grandpa**, Clyde W. Gabler
* **A History of First Fighter 1918-1983**, Charles F.O'Connell, Jr. Office of TAC History, Langley AFB, Virginia, June 1987
* The Squadron Diary, 71st Fighter Squadron, compiled by P.E. MacAllister
* Combat Squadrons of the Air Force, World War II
* Air Force Combat Units of World War II
* USAAF (Mediterranean Theater) Credits For The Destruction Of Enemy Aircraft In Air-To-Air Combat WW II Frank J. Olynyk 207 Chelmsford Dr. Aurora, OH 44202
* **The Lockheed P-38**, Warren M. Bodie, Widewing Publications
* **I Could Never Be So Lucky Again, An Autobiography**, Gen. James Doolittle, Bantam Books
* **Adorimini, A History Of The 82nd Fighter Group In WW II**, Steve Blake with John Stanaway, Walsworth Publishing Co., Marceline, Mo.
* **Messerschmitt Aces**, Walter Musciano, Tab/Aero Books, Blue Ridge Summit, Pa., 1989
* Missing AirCrew Reports, National Archives.

ITALIA

This narrative was handed around during the summer of 1944. No one seemed to know the origin let alone the author, however it well depicts wartime Italy.

If I were an artist with nothing to do
I'd paint me a picture, a composite view
Of historic Italia in which I would show
The variety of contrasts, the high and the low.
There'd be towering mountains, a deep blue sea
And filthy brats yelling Caramella! at me.
High plumed horses and colorful carts,
Two toned tresses on hustling tarts.
Chapels and churches great to behold,
Cash a king's ransom in glittering gold.
Poverty and want, no edible food,
Picked out of garbage practically nude.

Castles and palaces,opera house too
Hotel on a mountainside, magnificent view.
Homes made of bricks, clay, tile and mud
People all covered with scabs, scurvy and crud.
Palatial villas with palm trees tall,
A stinking apartment, mere hole in the wall.
Tree fringed gardens so swept by the breeze,
Goats wading in filth clear up to their knees.

Lovely cathedrals with perfect toned bells,
Mouldering catacombs, peace for the dead,
The living who noisily clamor for bread.
Creeping boulevards, a spangled team,
Alleys that wind like a dope fiend's dream.
Napoleonic caps, the Carabiniere,
Dejected old women with too much to carry.
A stately old gent with a Balboa beard,
Bare bottomed bambinos, both ends smeared.

Revealing statues all details complete,
A sensual figlia with sores on her feet.
Big breasted women, shabbily dressed
In discolored garments never been pressed.
Red peppers blooming all over a hill,
A sidewalk latrine, privacy nil.
Two-by-four shops with shelving quite bare,
Gesturing merchants arms waving in air.
Narrow gage sidewalks more like a shelf,
Butt puffing youngster scratching himself.

Lumbering carts hogging the road,
Nondescript trucks frequently towed.
Diminutive donkey, loaded for bear,
Horse drawn taxies seeking a fare.
Determined pedestrians courting disaster,
Walking in gutters where movement is faster.
Italian drivers all accident bound
Weaving and twisting to cover the ground.

Home made brooms, weeds tied to sticks,
Used on streets to clean off the bricks.
Bicycles and pushcarts blocking your path,
Street corner politicians needing a bath.
Barbers galore with manners quite mild,
Prolific women all heavy with child.

Mud smeared children clustering about
Filling their jugs from a common spout.
A dutiful mother with a look of despair
Picking the lice from her small daughter's hair.
Capable craftsmen, quite skilled in their art
In decrepit workshops, falling apart.
Intricate needlework out on display,
Surrounded by insects, rot and decay.
Elegant caskets carved out by hand
Odorous factories where leather is tanned.
Il Duce's secret weapon, kids by the score,
Caused by his bonus which ain't any more.

Arrogant wretches picking up snipes,
Miniature Fiats of various types.
Young street singers, hand organ tune,
Shoe shining urchins, a sidewalk saloon.
A beauteous maiden, a smile on her face
With a strong breath of garlic fouling the place.
Listless housewife, no shoes on her feet,
Washing and cooking right out in the street.

The family washing of tattletale gray
Hangs from the balcony blocking the way.
I'Tie native coffee, God what a mixture,
Tiled bathrooms with one extra fixture.
Families dining from one common bowl,
Next to a fish store, horrible hole.
Italian zoot suiters, all flashily dressed,
Barefooted beggars looking depressed.

Mud smeared children clustering about
Filling their jugs from a common spout.
A dutiful mother with a look of despair
Picking the lice from her small daughter's hair.
Capable craftsmen, quite skilled in their art
In decrepit workshops, falling apart.
Intricate needlework out on display,
Surrounded by insects, rot and decay.

A shoemaker's shop, a blackmarket store
Crawling with vermin, a beadscreen on the door.
I've tried to describe what I've frequently seen
I'm glad that I came but damned anxious to go.
Give it back to the natives, I'm ready to blow!

Transportation contrasts - 27th GI's en route to Rome for R & R. Monte Cassino in background. (Bob Share)

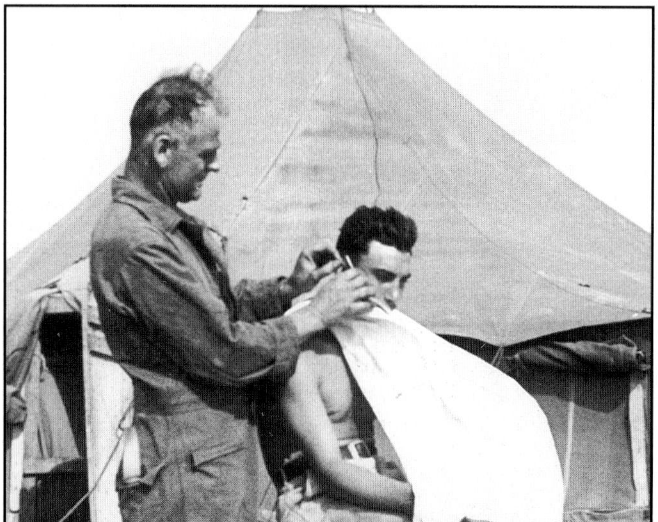

Time for a trim, Salsola, 1944. (Karpowicz)

Unloading belly tanks, Salsola. (Karpowicz)

Maj. Jerome Belleau 94th Exec. Officer. (Cake)

Advisory sign, 71st Club entrance, Salsola, 1944. (Hatch)

27th Maintenance and back-up crew at the Black Falcon Enlisted Club, Salsola. Front L - R Dunn, Key, Lange, Gall, Wingo, Van Hood, Lapinski, Anderson, Morris, Duby. Back L - R Furr, Toney, Clevenger, Wiggins, Christenson, Tanner, D. Anderson, Finkelstein, Fishgod, Meacham, O'Leary, Perryman, Krook, Johnson, Follett. (Caughlin)

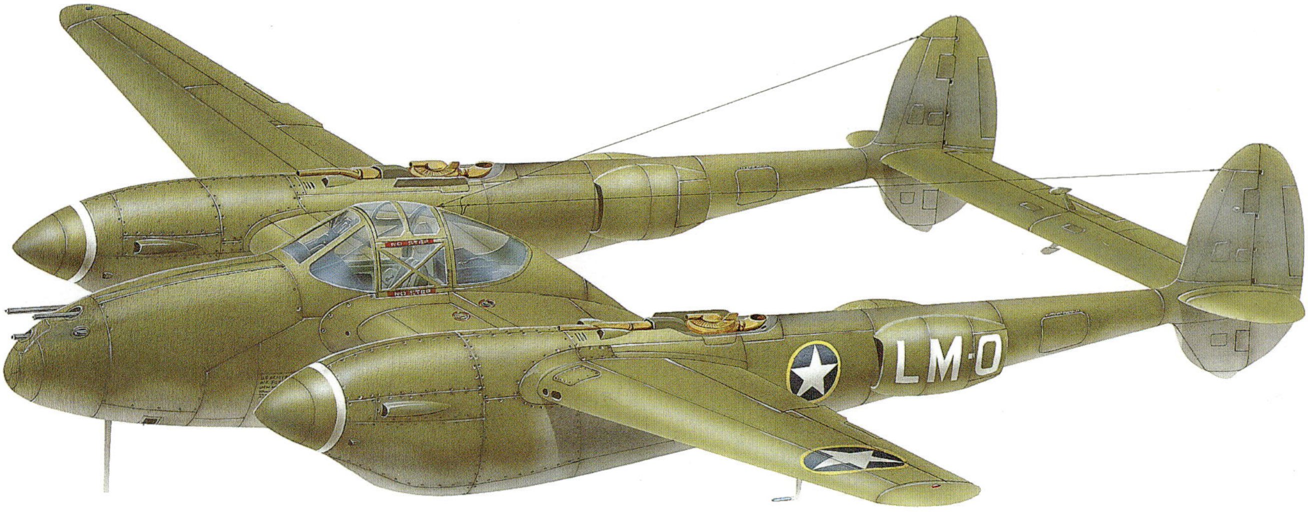

A 71st Fighter Squadron P-38F, serial no. unknown, is displayed in its distinctive Operation TORCH markings, the yellow ring around the national insignia, that was used by both U.S. and RAF aircraft during and shortly after the invasion of North Africa.

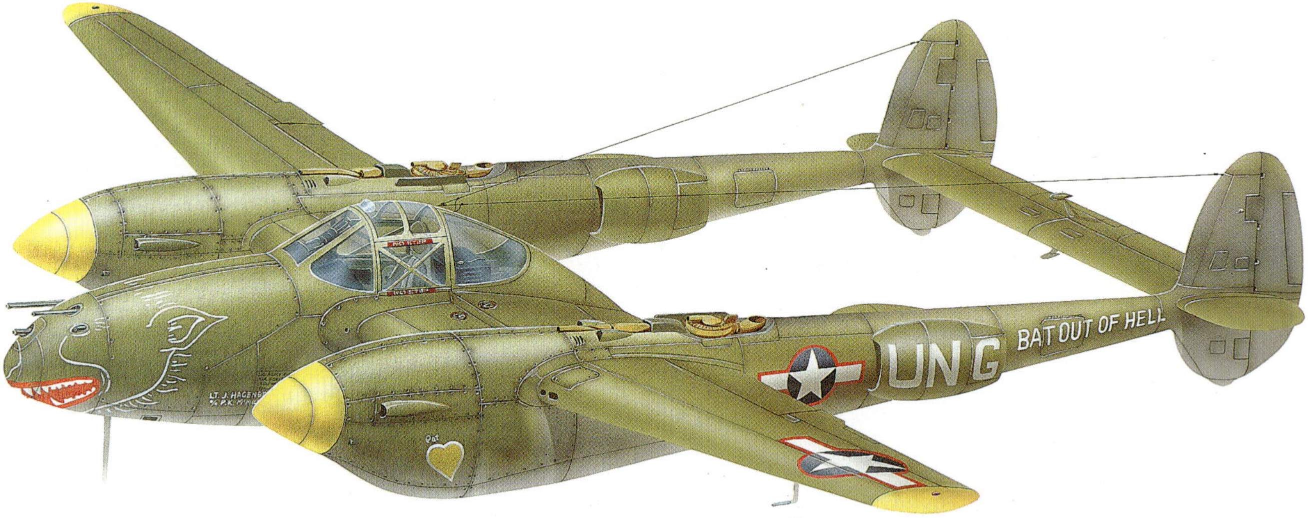

Captain Jim Hagenback, 94th Fighter Squadron, flew BAT OUT OF HELL a P-38F, serial no. unknown, shown here when national insignia was varied briefly in the summer of 1943 with a white bar and red surround.

Tom Maloney, 27th Fighter Squadron, flew several aircraft named MALONEY'S PONY. The last of these is depicted here, a P-38J-15, Serial No. 43-28746. He scored eight victories, one probable and three damaged in aerial combat.

Rumanian fighters encountered by the 1st Fighter Group near Ploesti on the disastrous mission of June 10, 1944, were Rumanian manufactured IAR 80C aircraft. This excellent fighter resembled the German Focke-Wulf 190 from a distance.

Right: 1st Fighter Group "Conquer or Die"

27th Fighter Squadron

71st Fighter Squadron

94th Fighter Squadron

27th pilots, Salsola, Top L - R Prout, Fleming, Ericson, Follet, Havard, Ganley, Huber, Lang, Rhoades, Rafael, Lawson, Enyeart, Hardin, Jones, Plummer, Jansen. Bottom L - R Erickson, Lowther, McHugh, Johnson, Kraft, Lau, Williams, Caughlin, Hanford, Pope, Lewis, William Johnson, Maloney, Kmonk, Adams. (Caughlin)

27th Squadron P-38's in echelon formation, summer 1944. (U.S. Air Force)

Sgt. Richard Abbott, crew chief with #23 MALONEY'S PONY, a P-38J-15 Salsola, September 1944. (Maloney)

94th pilots with awards. L - R Lee, Thiessen, Vrilakas, Sias. (Ilfrey)

Gordon Buttles, 94th pilot, shaving, Salsola, summer 1944. (Cake)

Warren Perryman, 27th nose art specialist, Salsola, 1944. Warren later designed 1st Ftr. Grp. memorial at Air Force Museum. (Bob Share)

Dan Darnell 94th pilot with Ron McEwen, Sq. S-2 on wing. (Lee)

John Mullins 94th pilot front left, Everett Jacobson, his crew chief front right 94th NCO club. Others are unknown.(Cake)

Salsola, Fall 1944 27th pilots L - R Pope (27th C.O.) Lang, Huber and Arnold. Capt. Huber has completed his missions. (Lang)

The 1st Fighter Group ground echelon disembarks at Cagliari, Sardinia. (Michener)

This P-38L-1, Serial No. 44-24184 was the aircraft of Lt. Gordon Buttles, 94th pilot. (Bostwick)

INDEX TO INDIVIDUALS
(Narrative and photo captions only)